THE DOMESTICATION OF WOMEN
Discrimination in Developing Societies

To my mother, Mary Katherine Rogers,
with love

The
Domestication
of Women
Discrimination
in Developing Societies

Barbara Rogers

ROUTLEDGE
London & New York

First published in 1980 in Great Britain
by Kogan Page Ltd.
and in the USA by
St. Martin's Press

First published as a Social Science Paperback in 1981
by Tavistock Publicastions Ltd
Reprinted in 1983

Reprinted 1989
by Routledge
11 New Fetter Lane
London EC4P 4EE
29 West 35th Street
New York, NY 10001

© 1980 Barbara Rogers

Printed in Great Britain by
J. W. Arrowsmith Ltd., Bristol BS3 2NT

All rights reserved. No part of this book may be
reprinted or reproduced or utilized in any form or by
any electronic, mechanical, or other means, now
known or hereafter invented, including photocopying
and recording, or in any information storage or
retrieval system, without permission in writing from
the publishers.

British Library Cataloguing in Publication Data
Rogers, Barbara
    The domestication of women
    1. Underdeveloped areas – Sex discrimination
    against women
    I. Title   **II.** Series
    305.4′2′091724   HQ1870.9

ISBN 0-415-04010-8

Acknowledgements

This book is the result of research done at the School of Development Studies, University of East Anglia, with the support of a two-year grant from the University for which I am most grateful. Raymond Apthorpe and Ingrid Palmer provided invaluable assistance and interest in the early stages, and Rayah Feldman and Bill Kinsey gave very helpful supervision and advice later. I also owe much to the United Nations Development Program for sponsoring my tour of agricultural projects in Africa, and especially to Erskine Childers and Paul Boyd for making it happen.

Help in the form of much vital material has been provided by friends and colleagues, of whom Adrienne Germain of the Ford Foundation and Lesley Yates, then of the Overseas Liaison Committee of the American Council on Education, have been particularly generous. I owe a debt also to many officials in the international agencies, especially Judith Bruce of the Population Council, Nancy Fee of the International Planned Parenthood Federation (IPPF), Natalie Hahn of the Food and Agriculture Organization (FAO), Elizabeth Reid of the United Nations, and Gloria Scott of the World Bank. Many Third World women gave advice and criticism, particular help being offered by Joséphine Gissou, Scholastique Kompaoré, Margaret Lubinda, Fatima Mernissi, Mrs Mogwe, Katherine Mwanamwamba and Zen Tadesse. Among international and government personnel in the field, Joy Greenidge, Bill Hereford, Tessa Lawrence, Celia Lotse, Mac Odell, Deborah Taylor and Sam Whittaker gave assistance far beyond the call of duty.

Since I have no wife to do my typing, I have relied heavily on the professional services of Sue Field Reid, Barbara Moxley and Tricia Newbury. I remain responsible, of course, for all the inadequacies of this book, and hope that it will nevertheless be a part of the process of discussion and action on the issue of discrimination against women in development planning.

Contents

Introduction

Women have been discovered — after a fashion. The development agencies are starting special projects for women all over the place, women's sections are set up in many countries, and there is even a United Nations Decade for Women with its own bureaucracy and series of conferences. Academics are engaged in profound debates about the 'status of women' and how it relates to their share of the work, while major questions remain unresolved about who should be speaking for Third World women.

At the same time, however, a survey of new books on development reveals that, apart from a handful of specialized books on women (of which this is one), development studies remain firmly orientated towards men, men being synonymous with all people. A search through the shelves of any library or bookshop specializing in development studies is likely to provide little or nothing by way of references to women and their work, the division of labour, gender roles generally, or even the euphemistic 'family labour'. In fact, the situation is even worse than it used to be, perhaps because of the new reliance on quantitative methods to provide an 'objective' view of economic development. In researching this topic, I was struck by the fact that for any reference to women's work in agriculture, by far the most important sector, it is necessary to go back to the colonial writers. And as this book will try to demonstrate, what is true of the development studies literature is also true of development planning.

This study is not primarily about women, but about how planners relate to them. At a number of meetings, Third World women have made it clear that they are the ones who will analyze their own situation. For Westerners, a more useful contribution is to tackle the problem of the discriminatory processes in their own culture and society, which are being imposed on others through what we like to call development. This book is therefore about planners trained in the Western tradition, and the impact of their received ideas about women in general on poor women in the Third World. I am using the word 'planner' in a very broad sense to include all those professionals and 'experts' at the decision-making levels of international and bilateral aid agencies, at headquarters and in the field, and national and local officials in the Third World: all those, in fact, who determine the formulation, design and execution of development policies, programs and projects. Almost all of these planners, for reasons discussed in the book, are men. One of the village women I met in Zambia summed up the problem in demanding that I convey her message to these 'big men': the women had done everything they could to help themselves and support each other, and had raised some money by their own efforts although not nearly enough for what they wanted. It was time the planners paid some attention to the women's needs.

The work presented here is in three parts. The first discusses Western male ideology about gender distinctions and the division of labour, and how their interpretations of other societies are used to bolster myths about women's 'natural' place in society. Secondly, the planning process itself is analyzed: discrimination against women in the development agencies, the distortions involved in the research and data collection on which development planning is based, and the relegation of Third World women to special projects in the 'domestic' ghetto. Finally, the discriminatory impact of the planning process is outlined in terms of subsistence agriculture, the sector in which most Third World women are concentrated. The general trends as regards the impact of development planning on women's access to resources and on their work-load are discussed briefly, together with the implications of women's problems for the success or otherwise of the whole development process.

Part One:
Problems of Perception

Introduction to Part One

The assumptions that development planners make about women in society are almost never stated, but are all the more powerful for that reason. It is thought 'natural' that a woman's place is in the home and that she has a very specific set of tasks which are thought to be universal because they are based on the biological imperatives of sex. The most important role for women, defining their entire life, is portrayed as the bearing and bringing-up of children. A man, on the other hand, is seen as the 'natural' head of the family, its representative in the outside world, and therefore the person with whom planners will deal. Since it is assumed that men control families ('the master in his own home'), any new resources intended for everyone should logically be channelled through them.

Chapter 1 discusses some of the misconceptions held by development planners, as products of the Western tradition, about women and how they relate to men. We review in Chapter 1 some questions about sex and gender, the division of labour, the role of women and men in child-care and the role of women as domestics in modern Western society. In Chapter 2 the definition of the problem in terms of the 'status of women', as it has come to be called, is scrutinized. The ways in which Westerners describe women in what they call 'primitive societies' are also questioned. The processes of discrimination involved in development, with its dual economy relegating women increasingly to the 'underdeveloped' subsistence sector, is presented as a means of understanding what is happening in the Third World today.

Chapter One

Women and men:
the division of labour

Sex and gender

Sex is a physical distinction; gender is social and cultural. Although masculine or feminine gender is usually associated with male or female sex, this is not an absolute correlation. In Western society a child is conscious of the gender of its upbringing well before it can talk properly, and any attempts to change the gender of rearing because of 'mistaken' sex-labelling (biological characteristics at birth can often be ambiguous) frequently results in severe disturbance after the age of two. Many transsexuals are unambiguously of one sex, but identify themselves as of the opposite gender. A comprehensive survey by Ann Oakley has helped to clarify the important distinction between biological sex and the enormous range of distinctions made by Western society in the name of gender — feminine and masculine.[1]

Gender and the division of labour

In talking of the division of labour between women and men in different societies we are talking almost exclusively of gender roles rather than sex roles, determined by culture rather than biology. Virtually all human behaviour, including even such 'physical' activities as copulation, childbirth and the parental care without which children cannot survive, is learned behaviour. It varies widely among different societies, and is quite distinct from that of the lower animals where activities are conducted, to a much greater degree, without prior experience or learning. It has been argued by Clifford Geertz that people (whom he calls 'man') are unique in that they are not highly programmed and do not perform actions basic to survival through intrinsic processes, but need to learn.

Geertz stresses that a major fact about our central nervous system is 'the relative incompleteness with which, acting within the confines of autogenous patterns alone, it is able to specify behaviour.' Thus culture is of crucial importance in, among other things, determining the role of gender. This applies even to sexual activities and to the bearing and rearing of children, as well as to other distinctions less directly linked to biological sex.[2]

Gender-roles, then, are determined to a relatively small extent by sexual characteristics. However, we have to account for the fact that the division of labour by gender is a factor in most, if not all societies. It plays a role in the production process — although the lines of demarcation show almost infinite variation and some societies will have more strictly defined areas of 'female' and 'male' activity than others. Gender is, together with age, a widely used means by which societies make some form of division of labour, a process of specialization which is an important tool of efficiency in any production system. There are many other criteria which are also used, such as racial or other physical distinctions, geographical origin, 'family' or 'tribe' and so on, and also to some extent caste or class. These are all cultural interpretations of physical differences. Margaret Mead, who did pioneering work in revealing the wide range of psychological and cultural traits which can be attached to masculine and feminine gender-roles, found that important characteristics of women in one culture were often those of men in another.[3] She concludes from her field-work:

> 'Primitive materials, therefore, give no support to the theory that there is a "natural" connexion between conditions of human gestation and appropriate cultural practices.'[4]

There is little physical determinism about which gender performs which role, apart from pregnancy and childbirth (and even these may be elaborately simulated by the father, who would then receive the credit for it).[5] Even where Westerners commonly assume that only women would participate in a particular activity — for example breast-feeding — the pattern is by no means as clear-cut as the Western model would have us believe. Margaret Mead, in an attack on the sexual determinism advocated by Dr John Bowlby and others, promoting the mother as the only person able to care for a baby, has pointed out that in many 'primitive' societies both men and women — including those who have never given birth — may give the breast to a small child, and may offer it premasticated food right from birth.[6] She lists several examples of societies which insist on the men being primarily responsible for infant and child-care.[7] When Manus children were offered dolls for the first time it was the boys who took and played with them,[8] while in other societies this would be seen as consistent with feminine roles.

Women's work

The actual pattern of female and male activities will be devised by each society according to its beliefs about the reproductive functions of the sexes. Where many Western societies see rest as appropriate for women in pregnancy and labour, others require hard work and exercise. Childbirth has completely different and incompatible meanings, and is a different kind of event, according to the society in question.[9] Western societies see men as largely superfluous in pregnancy and childbirth; other societies believe them to be vitally concerned, and restrict their activities during their wives' pregnancy. Men may also simulate labour and be treated accordingly;[10] and they may play a large part in nurturing children almost from birth.[11] Bronislaw Malinowski observed that among Australian aborigines, if a child's father dies before it is born the child is killed by the mother. Eskimos of Greenland kill a baby if its mother dies in childbirth, for the same reason.[12] Either variant makes equal sense in terms of the gender which has been socialized in what we call the 'maternal' role.

The reproductive and nurturing roles, however they are assigned by the culture between the two genders, may serve to define broad lines of division among new tasks. George Murdock explains it by saying: 'New tasks as they arise are assigned to one sphere of activities or the other in accordance with convenience and precedent.'[13] These are of course matters of culture and the individual interpretation of culture.

In Western industrial society gender distinctions are commonly rationalized by beliefs about the central importance of women's role in child-rearing, and the imputed operation of a maternal 'instinct'. There is also the assumption that all men are 'naturally' incapable of nurturing children and, to compensate, are 'naturally' stronger than all women, who are deemed incapable of heavy work. The work that women perform, regardless of its actual character, is seen as somehow 'not-work', or at best very light work. In many other societies, the reverse beliefs apply. Malinowski, despite believing implicitly in the Western prescription as to the 'natural' characteristics of women, sums up his observations in *The Sexual Life of Savages*:

> 'It is easy to see that the amount of work allotted to women is *considerably greater* and that their labour is much *harder* than men's

work . . . . Heavier work ought naturally to be performed by men; here the contrary obtains.'[14]

In an attempt to quantify the incidence of various patterns of division of labour by gender, George Murdock surveyed the available data for 224 societies, most of them non-literate. With a list of 46 different activities, he suggests that some are more often masculine than feminine, and some the reverse. For example, lumbering is exclusively masculine in 104 and exclusively feminine in six cases; cooking is exclusively feminine in 158 and exclusively masculine in five. Hunting, fishing, weapon-making, boat-building and mining tend to be masculine; grinding grain and carrying water tend to be feminine. Activities, that are less consistently allotted to one gender include preparing the soil, planting, tending and harvesting the crops and burden-bearing.[15] This kind of survey, based on very crude data often not intended for this kind of analysis and with a relatively small number of societies being studied, is of rather limited value. However, it serves to indicate that even activities which are usually exclusively masculine or exclusively feminine do not invariably have to be so. One occupation sometimes assumed to be always for men only is war; but Ann Oakley is able to list various instances, in modern as well as ancient times, in which women have taken part in war.[16] At the time of writing there are women fighting with liberation movements in Zimbabwe, Namibia, Western Sahara and Eritrea, and with urban guerrilla movements in various countries.

How sharply the distinction is made between feminine and masculine roles of course varies within a wide range. In some cultures, a special category may be created for women who excel in pursuits assigned to both sexes.[17] In parts of Africa wealthier women can even acquire wives to perform women's work in their own and their husband's fields and compounds: they will pay the bride-price and go through a normal marriage ceremony.[18] In many societies the division of labour is flexible. Margaret Mead points out that where women withdraw periodically from their families — a common enough feature of many societies — men have to do the housework and child-care even if the women do that for the rest of the time.[19] Among the Bemba of Zambia, men would normally sew, wash clothing and sometimes cook, but would have to take over all the women's tasks as well when the chiefs required tributary labour.[20]

Sometimes the division of labour is merely preferential, and very little anxiety is shown by either gender over temporary reversals of the guidelines. Cora Du Bois reports that in Alor it is not thought unhealthy for anyone to take on the other gen- der's work; in fact, those who do, are admired for possessing a supplementary skill. Many men are passionate horticulturalists (this being feminine work) and many women have financial skills (masculine work).[21] The Mbuti pygmies have been des- cribed by Colin Turnbull as having a social structure in which sex- or gender-roles seem to be negligible. Both genders take part in the main activities of hunting and gathering. They also share political decisions and have the same social 'status'. There is very little division of labour by gender; for example, men as well as women often care for even the youngest children and pregnancy is no bar to hunting. The Mbuti language distinguishes by gender only in terms of parenthood, ie 'mother' and 'father'. Where other societies in their rituals emphasize the distinctions between the genders, Mbuti rituals emphasize the lack of them. There seems to be little anxiety about gender-roles.[22] Margaret Mead also found a lack of differentiation among the Arapesh of New Guinea.[23] Geertz notes a similar pattern in Bali:

> 'Sexual differentiation is culturally extremely played down in Bali and most activities, formal and informal, involve the participation of men and women on equal ground, commonly known as linked couples. From religion, to politics, to economics, to kinship, to dress, Bali is a rather "unisex" society, a fact both its customs and its symbolism clearly express.'[24]

At the other end of the scale are societies which impose ex- tremely rigid gender-roles. In some cultures where horticulture, for example, is defined as women's work, a proclivity for it in a man is regarded as proof of sexual deviation.

One example of a society which imposes extremely rigid gender-roles is that of the Mundurucú Indians of central Brazil, where the polarization of gender-roles and groupings is a prim- ary structural element; physical and social separation is virtu- ally complete. Each gender-group interacts almost entirely with itself, and antagonism between the two is shown on many ritual and other occasions — for example, the practice of gang rape by men on women who are seen as deviating from their strictly defined role. The gender distinctions pervade the area of person- ality, and anxiety is expressed in many ways about even a tem- porary reversal of gender roles, apparently a stronger factor for men than for women.[25] Ann Oakley suggests that this anxiety is most likely to be found in societies where gender is a major organizing principle of social structure.[26] This is certainly the case in modern Western society.

It is necessary here to remind ourselves that gender distinctions are always modified by a wide range of other social distinctions, such as age. The relative importance of gender will vary from society to society as well as operating in different ways according to whether, for example, the two genders live largely segregated from each other or highly integrated; the relative value placed on feminine and masculine roles; the importance of age and wealth as they modify gender roles, together with many other variables. Yolanda and Robert Murphy are themselves careful to point out in their study of the Mundurucú that, although gender roles are among the most basic forms of social distinction, it would be impossible to understand the content of these roles and their articulation with each other without reference to kinship and how it overlaps with neighbourliness and friendship.[27] For example, older women behave much more like men than younger women. The ways in which individuals deal with each other depend very much on their kinship which will determine how gender roles come into play. The relationship between a woman and a man of the same moiety would be very different from that between a woman and a man of different moieties who are eligible to marry each other.[28]

Women in Western Society

In examining the impact of Western gender-distinctions on pre-colonial societies through the process of colonial and 'development' policy, it is essential to distinguish between what we generally believe our own practice to be, and what it in fact is. With the increasingly complex division of labour in industrial society, the basic divisions of gender and age have lost their original functions as they operated in a peasant society, as a function mainly of kinship. Gender distinctions, and the interaction of women and men in the family are now based on their respective economic relations outside the family, and the position assigned to them by social class, education and other external factors. The beliefs about differences between women and men, which usually refer back to some mythical prehistorical society where men always hunted and women always stayed in the cave with the children, remain a very important element in social and cultural organization.

The idea that Western gender roles are the 'natural' ones is used in support of a male ideology which seeks to exclude women from many important areas of modern life. In comparison with other societies, Western industrial society maintains a high degree of segregation in the sphere of work, while there is, at the same time, an equally high degree of integration of women and men in residential patterns. In anthropological terms, families are based on patrilineal and virilocal organization which is justified as 'natural'. The patrilineal system means that children are identified as belonging to their fathers' line, as indicated by the use of the father's surname, and children without properly recognized fathers are stigmatized as 'illegitimate'. Residence is almost always determined by the husbands' interests in terms of work and social networks, and can therefore be termed 'virilocal'. Women are classed as 'dependants' of their husbands, particularly for financial purposes — as seen for example in the tax laws — and are treated in some ways like their children. For example, in many cases they need their husband's permission to obtain certain kinds of medical treatment, particularly anything relating to their fertility (thus, married women usually cannot be sterilized without their husband's consent, although men can be vasectomized without their wives' consent).

The social class and lifestyle of a family are determined mainly by the husband's occupation outside the family and he is expected to control what happens inside it, even to the extent of deciding whether or not 'his' wife should have a paid job outside. Any deviation from this is seen as threatening the man's very identity: for example, it is believed to be shameful for a husband to earn less than his wife; to be less tall; for the wife to have a stronger personality or greater intelligence. While primarily an ideology of marriage, this applies to almost all relations between women and men. Girls learn to fear success and to underachieve, concealing intelligence and initiative in order to be accepted as 'feminine'. Men are expected to be 'aggressive' and unemotional, women to be sensitive, intuitive etc. From a very early age they learn what is expected of them in terms of the 'feminine' or 'masculine' personality, and this is heavily reinforced at puberty.[29]

In psychological terms, considerable effort and anxiety are associated with the precise definition of 'feminine' and 'masculine' gender characteristics and functions, more particularly perhaps by men for whom the ideology of manhood or virility can impose serious emotional problems.[30] It can also provoke great anxiety and hostility if gender roles as such are questioned by women. The extreme differentiation of gender-roles and personalities is also associated, according to Anthony Storr and others, with a great deal of marital conflict.[31] There is considerable latent or overt hostility between the genders,[32] and a high degree of violence shown by men toward women, particularly in the form of sexual assault and domestic violence. These forms of attack are viewed very tolerantly by men as a whole, in comparison with other forms of violence for which society reserves serious sanctions. Rape victims are treated unsympathetically by the police and if the case comes to court it is the victims who are put on trial as regards their credibility and sexual experience. 'Wife-beating' is regarded as a trivial offence, the police and the courts often refusing to intervene. There is no law against men soliciting sex from female strangers, although women who do this to male strangers are labelled as common prostitutes and liable to penal sanctions. Similarly, there is no social sanction against men's harassment of women, especially the younger ones, on the streets and in public transport. Women are continually confronted with degrading advertisements and pornography which publicly present them as sex objects, often inviting violence. Many men find any protest about women's feelings on sexual harassment and pornography a matter for ridicule, and they refuse to accept that it happens or that women's feelings about it have any significance.

The attitudes of individual men will vary enormously according to the rank (by class, ethnic group, appearance, age etc) of an individual woman, ranging from paternalism, perhaps manifested by 'gallantry', to indifference or extreme hostility. Attitudes of men as a whole can also change very rapidly according to convenience: thus the belief in women's 'frailty' is discarded when they are needed in heavy industrial or agricultural work because of major wars, only to be renewed with a barrage of propaganda about a women's place being in the home, when the men want the jobs back, and then the 'maternal deprivation' of children whose mothers dare to leave them for even a short time is stressed.

There is no one immutable pattern of gender-roles that can be presented as the norm for Western industrial society, and which is imposed on other cultures as a blueprint for development. Rather, there is a process, a prolonged struggle between women and men for control of their own and each other's activities and attitudes. It is this conflict which has shaped the division of labour and the ideology of women's place. Radcliffe-Brown suggested in 1930 that there was some 'law of opposition' involving:

' . . . some form of socially regulated and organized antagonism, between [a group] and other groups or segments with which it is in
contact, which opposition serves to keep the separate segments
differentiated and distinct.'[33]

The stress on this antagonism between groups as an element
'socially regulated and organized' seems, however, too well-
ordered to account for the gender-role differentiation at work
in the process of development. More convincing is Van Velsen's
'situational analysis' approach, proposed as being particularly
suitable for the study of unstable and non-homogeneous societies, in which conflict is seen as a 'normal' part of social process. It looks squarely at the discrepancy between people's
beliefs and professed acceptance of certain norms on the one
hand, and their actual behaviour on the other, recognizing that
disparate systems of belief co-exist, and are called into action
in different social situations.[34]

The description by Regina Morantz of 19th-century American gender-roles — which could quite easily be adapted to the
20th century — illustrates well the appositeness of situational
analysis to this problem:

'Woman's image was riddled with contradictions; guardians of the
race, but wholly subject to male authority; preserver of civilization,
religion and culture, yet considered the intellectual inferior of men;
the primary socializer of her children, but given no more real responsibility and dignity than a child herself.'[35]

'Feminine' roles in Western society: housewife and mother

Perhaps the most striking feature of Western male ideology is
the enormous emphasis on the exclusive role of the biological
mother (or another individual woman as mother-substitute) in
nurturing infants and children, particularly in the first few
years. This is closely linked with the identification of women's
place as the domestic sphere, as wives and mothers: the home is
presented to them as their primary occupation even if they take
a second, salaried job outside the home. There are important
psychological processes involved in the socialization of girls in
contemporary society which makes many women into vigorous
advocates of their home-and-family destiny; they have been
carefully trained for this work, and are repeatedly told that it
represents the best that life can offer.[36] The woman who
refuses to allow her husband into the kitchen is claiming control
at least of the area of work she has been encouraged to call her
own.

It is a common belief that, over recent decades, women in general have been 'liberated' from the domestic drudgery of the past by the introduction of domestic appliances. In fact, the work has merely changed in character: new standards and expectations, combined with the increasingly exclusive responsibility of adult women for a family's housework — as children attend school full-time until adulthood, and extended families are broken up — have resulted in women working similar or even longer hours to those prevailing before the introduction of domestic appliances and convenience foods. They also now work much longer than men. A time study of housework in the 1930s showed that it took an average of 61 hours a week in farm households, which lacked most of the modern appliances, as compared with 81 hours a week in large cities, where the greatest concentration of these appliances was to be found. This contrasted with strong moves in paid employment for a 40-hour working week.[37]

Conditions of work have also deteriorated. Anxiety and depression among women, to a degree greater than in the past, are provoked by new forms of housework and these are now a commonplace of modern medical practice. The isolation resulting from the fragmentation of households into nuclear families and the new burden of exclusive full-time child care is combined with constant and nerve-racking vigilance required to protect small children from the dangerous chemical, electrical and other hazards of modern housing.[38] The whole new ideology of 'maternal deprivation', which isolates women with their small children for several years at a time, is also a powerful instrument in isolating 'housewives' while inducing anxiety and guilt.

Lenin's description of women's primary job, emphasizing its monotony and lack of job satisfaction, holds good now as much as it did in his time:

> 'Housework is the most unproductive, the most barbarous and the most arduous work a woman can do. It is exceptionally petty and does not include anything that would in any way promote the development of the woman'.[39]

In a rare study of British housewives in the Depression, Margery Spring Rice emphasized the enormous burden placed on women by their 'housewife' role:

> '. . . in the large majority of homes, the woman is . . . the slave without whose labour the whole structure of the family tends to collapse.'[40]

This work is essential to our social and economic structure of which the family is the basic unit; at the same time it is petty, isolated, monotonous and requires virtually unending hours of work, much of it hard and unrewarding. It is highly labour-intensive; but at the same time it is unpaid. Housewives have very little disposable income apart from the housekeeping money and they frequently have to account to their husbands in detail for any personal expenditure.

It is important to understand the nature of women's domestic work in Western industrial society because it is all too readily believed that women 'do not work' as long as they do housework full-time. Housewives themselves, although often resentful of the charge that they 'do not work' will demonstrate shame at their lack of a paid job: 'I'm only a housewife'. It is no accident that women are largely confined to housework, when its social prestige, as well as pay and job satisfaction, are at the bottom of the work hierarchy. This is the result of a process which has been traced back to the 11th century in Europe, and more importantly perhaps to the industrial revolution, whereby women lost their economic autonomy as producers in their own right — as farmers, craft workers or traders — and became increasingly dependent on the wages of men. At the same time, they became more confined to domestic or 'house' work. And the modern cash economy became increasingly divorced from the domestic or subsistence economy.[41]

Domestication

The process of confining women to the domestic sphere was applauded by many 'progressive' social thinkers of the industrial revolution period. Auguste Comte, the apostle of progress and rationality, justified the process of domestication of women by the supposed need for men 'to assure [woman's] emotional destiny' by making her life 'more and more domestic' and 'above all detach her from all outside work . . .'[42] It is this separation out of the modern economy, with the central role of paid work from the unpaid 'domestic' or subsistence sector, which is critical also in the Third World development process. Of course the manifestations of this domestication (except for a small urban élite following the Western pattern) are at this stage very different for Third World women than they are for Westerners.

Urbanization and industrialization have been associated — particularly as men and their trade unions began to get women out of many jobs through 'protective' legislation — with the phenomenon of the 'domestic science' movement which sought to provide a 'scientific' rationale for confining women to unpaid domestic work. The domestic science movement of the early 20th century, although nominally led by individual women, was backed by powerful vested interests in the male order: men's clubs, breweries and other powerful donors and supporters.[43] The American Medical Association, for example, joined the campaign for 'domestic science' education in 1899 on the grounds that it would lead to a reduction in 'divorce, insanity, pauperism, competition of labour between the sexes, men's and women's clubs, etc'.[44]

Barbara Ehrenreich and Deirdre English, describing the American scene, have illustrated how the movement promoted 'home-making' as a vocation justifying the unemployment of women, in order to serve men. Teachings were based on the concept of 'scientific' housework and domestic management: 'the home is the social workshop for the making of men', as Henrietta Goodrich told a domestic science gathering in 1902.[45] This idea had to be introduced to the poor immigrants of the urban slums; a variety of pseudo-scientific theories were advanced through missionary-type work, and domestic science lessons were used to make girls dissatisfied with their homes and aspiring to a more 'ideal' domestic life requiring increased levels of consumption.[46]

The scientific basis of 'domestic science' is weak. Ehrenreich and English point out that the frenzy of cleaning and dusting imposed as a moral imperative on American women rested on an extremely dubious 'germ theory'; failure to maintain everything free of dust and germs was suggested as being 'akin to murder'. Yet the scientific content of 'scientific cleaning' has always been extremely small. For example, the pioneering domestic scientists believed that the major household germ carrier was dust, which has since been completely disproved. Various 'scientific' cleaning practices were promoted which had no scientific basis whatever (although they are still widely practised). There is, in fact, no evidence that frequent house cleaning has any relation to the health of the occupants — except that some, especially the cleaner, may be allergic to the dust stirred up in this way. Similarly, Ehrenreich and English describe a variety of domestic tasks which amounted to the manufacture of work with no particular benefit.[47]

Another major factor in the domestication of Western women has been the idea of 'maternal deprivation' from which children would allegedly suffer if their mother, or another woman working full-time as mother substitute, did not provide them with 'constant attention night and day, seven days a week and 365 days in the year',[48] as instructed by John Bowlby and others, supported by massive publicity from the end of the Second World War and continuing to this day. The function of this ideology has been widely observed as forcing women out of their war-time jobs in industry and elsewhere and into unpaid domestic work, with constant appeals to their feelings of guilt about permanent traumas which they would allegedly give their children if they left them, especially for jobs outside the home.

Like domestic science, maternal deprivation theories have a very weak basis in terms of scientific research. Barry Hill calls Bowlby's ideas 'folklore'[49] and Lee Comer describes them as an 'odd mixture of sentimentality, dubious logic and straight-forward prejudice'.[50] The suggestion that 'maternal deprivation', arising out of women working outside the home while their children are small, as the basic cause of juvenile delinquency has been attacked by many experts in the field.[51] Bowlby's arguments rely heavily on the disturbances shown by children in institutions and refugee camps, with an illogical leap from that to statements that the absence of a full-time mother is the sole cause of any such disturbance.[52] Studies of delinquent boys, by other investigators, have indicated that the major common feature in their family environment was actually the lack of an effective father: it is primarily in a loving relationship with him that a boy can successfully learn his role in our society.[53] The myth of 'motherhood' as the critical, in fact the only influence on early childhood development, has been described by Ann Oakley elsewhere as contributing to children's maladjustment, since it is the validating myth for the denial of paternal responsibility, especially in giving love and care to infants and children.[54]

Margaret Mead stresses the hazards to small children of being brought up to rely on a single parent-figure like the mother; apart from the limitations on their development they are badly affected by even temporary separations from that person whereas other children would have greater security and continuity of care where it is provided by a variety of people.[55]

Domestication and discrimination

The domestication of women in terms of housework and child-care is closely linked with various forms of discrimination against

women in the non-domestic sphere. Thus the restriction of educational opportunity is rationalized in terms of the future
domestic career as wives and mothers that awaits all girls. In
hiring, promotion and wage structures, the arguments are all
similar, with the added rider that women do not need to 'work'
(ie be paid for their work) because they are supposed to have
husbands supporting them and their children. These arguments
are of course invalid in the case of the majority of women
seeking paid employment: in the United States, for example,
two-thirds of all employed women have no husband, while in
most cases they have many children and other dependants of
their own; or if they are married, their husbands earn less than
the officially defined poverty level income for families.[56] However, the lack of foundation in fact for the beliefs about women
not needing to receive a wage for their work, does not detract
from the social force or function of these beliefs. They reinforce
the segregation of women into unpaid domestic (or subsistence)
work, or into the lowest paid and least upwardly mobile ranks
of paid labour.

The new division of labour, which identifies women with the
domestic sphere and men with the 'outside world' of the
modern economy, is both cause and effect of the virtual monopoly by men of the important positions in the socio-economic
hierarchy, and their associated control of the main institutions
of modern society: law, politics, public administration, male
clubs, the armed forces and police, commerce, industry and
banking, trade unions, the media, and other major institutions.
The relationship between women and men has indeed become
increasingly subject to men's use of their position in the political, legal, academic and economic institutions.

A phenomenon of recent decades has been a tendency in
many countries for the number of women employed to fall;[57]
a far smaller proportion of West European women now do paid
work, especially full-time paid work, than at the beginning of
the 20th century. A continuing characteristic of women in paid
employment is the insecure nature of their jobs: at times of
recession in particular they are the first to be fired, and they
suffer more from unemployment than men.[58] Women's work is
also paid considerably less than men's, even when the end product is the same. In Britain, the gap between the bottom 90 per
cent of women's wages and the corresponding male wages was
greater in 1959 than in 1911.[59] A substantial and increasing
proportion of women are employed in the notoriously low-paid
home-working and part-time sectors.

Yet while job opportunities and conditions of work remain as bad, or even worse than they have ever been for most women, there is a prevailing belief that women are taking jobs for the first time in history, and that they are earning wages much closer to those available to men. This belief is fostered by a few highly publicized cases of individual women, together with the token activities of bodies like the Equal Opportunities Commission. Cases taken to industrial tribunals under the Sex Discrimination Act or Equal Pay Act have often been treated with derision by male judges, and these laws are continually flouted even by the Government, in Britain as elsewhere: the manpower training schemes, for example, are almost exactly what their name implies.

As a low-level reserve labour force women show many parallels with blacks. For example, white men as a group will commonly show hostility towards women or blacks being hired or promoted over them, threatening their position as the élite labour force; the trade unions are an important instrument in this.[60] Job discrimination is one important manifestation of the overall lack of organized power of women, which operates in a similar way to discrimination against black people or Jews.[61]

References

1. Ann Oakley, *Sex, gender and society*, London, Temple Smith, 1972.
2. Clifford Geertz, *The interpretation of cultures*, London, Hutchinson, 1975, pp 91-92 and ff.
3. Margaret Mead, *Male and female: a study of the sexes in a changing world*, Harmondsworth, Penguin Books, 1962, reprinted 1975.
4. Cited in Oakley, *Sex, gender, op cit*, p 147.
5. See for example Ernestine Friedl, *Women and men: an anthropologist's view*, New York, Holt, Rinehart and Winston, Basic Anthropology Units, 1975.
6. Margaret Mead, 'A cultural anthropological approach to maternal deprivation,' in World Health Organization, *Deprivation of maternal care: a reassessment of its effects*, Geneva, WHO, 1962, p 56.
7. Cited in Ann Oakley, *Housewife*, Harmondsworth, Penguin, 1976, chapters 2, 7.
8. Cited in Oakley, *Sex, gender, op cit*, p 196.
9. Mead, *Male and female, op cit*, p 222.
10. Bruno Bettelheim has an extensive account of rituals symbolizing what he regards as 'womb envy' in non-literate societies, in *Symbolic wounds: puberty rites and the envious male*, New York, The Free Press, 1954.
11. Oakley, *Sex, gender, op cit*, pp 131-32, 134-37.
12. Cited in *ibid*, pp 135-36.
13. Cited in *ibid*, p 146.
14. B Malinowski, *The sexual life of savages*, London, Routledge, 1932.

15. G P Murdock, 'Comparative data on the division of labour by sex,' *Social forces*, Vol 15, No 4, 1937, pp 551-53.
16. Oakley, *Sex, gender, op cit*, pp 145-46.
17. Oakley, *Sex, gender, op cit*, pp 147-49.
18. Friedl, *Women and men, op cit*, pp 65 and ff.
19. Margaret Mead, *Male and female, op cit*, pp 164-65.
20. Friedl, *Women and men, op cit*, p 118.
21. C Du Bois, *The people of Alor*, University of Minnesota Press, 1944.
22. Colin Turnbull, *Wayward servants*, London, Eyre and Spottiswoode, 1965.
23. Mead, *Male and female, op cit*, pp 164-65.
24. Geertz, *Interpretation of cultures, op cit*, pp 376-77, 417-18.
25. Yolanda Murphy and Robert F Murphy, *Women of the forest*, New York, Columbia University Press, 1974.
26. Oakley, *Sex, gender, op cit*, pp 149-50.
27. Murphy and Murphy, *op cit*, pp 70 and ff.
28. *Ibid, passim*.
29. A substantial literature has appeared on these processes in recent years; see eg Martha Tamara Shuch Mednick, Sandra Schwartz Tangri and Lois Wladis Hoffman (eds), *Women and achievement: social and motivational analyses*, Washington DC, Hemisphere Publishing Corporation, and New York, John Wiley and Sons, 1974, sponsored by the Society for the Psychological Study of Social Issues.
30. For overviews see Oakley, *Sex, gender, op cit*, pp 149-50; and Patrick C Lee and Robert Sussman Stewart (eds), *Sex differences: cultural and developmental dimensions*, New York, Urizen Books, 1976.
31. Oakley, *Sex, gender, op cit*, pp 149-50.
32. See eg Ann Whitehead, 'Sexual antagonism in Herefordshire,' in Diana Leonard Barker and Sheila Allen (eds), *Dependence and exploitation in work and marriage*, London, Longman, 1976.
33. D F Pocock, *Social anthropology*, London, Steed and Ward, 1961, pp 77-78.
34. J Van Velsen, 'The extended-case method and situational analysis', in A L Epstein (ed), *The craft of social anthropology*, London, Tavistock Publications, 1967.
35. Regina Morantz, 'The lady and her physician', in Mary Hartman and Lois W Banner (eds), *Clio's consciousness raised: new perspectives on the history of women*, New York, Harper and Row, Harper Torchbooks, 1974, p 38.
36. See eg Mednick *et al, op cit*.
37. Cited in Mead, *Male and female, op cit*, p 355.
38. Lee Comer, *Wedlocked women*, Leeds, Feminist Books, 1974, pp 97-98.
39. V I Lenin, cited in *ibid*, p 83. The nature of housework is described in Comer and also in Oakley, *Housewife, op cit*.
40. Margery Spring Rice, *Working class wives: their health and conditions*, Harmondsworth, Penguin, 1939, p 14.
41. See eg Evelyn Sullerot, *Women, society and change* (translated from the French by Margaret Scotford Archer), London, Weidenfeld and Nicolson, World University Library, 1971; also Elizabeth Janeway, *Man's world, woman's place: a study in social mythology*, Harmondsworth, Penguin, 1977; and JoAnn McNamara and Suzanne Wemple, 'The power of women through the family in medieval Europe: 500-1100', in Hartman and Banner (eds), *op cit*, pp 103-118.

42. Auguste Comte, *Discours sur l'ensemble du positivisme*, Paris, Mathias, 1848, pp 91, 204, 243-53.
43. See Brian Harrison, *Separate spheres: the opposition to women's suffrage in Britain*, London, Croom Helm, 1978.
44. Editorial, 'Public school instruction in cooking'. *Journal of the American Medical Association*, Vol 32, 1899, p 1183.
45. *Proceedings of the Fourth Annual Conference on Home Economics*, Lake Placid, New York, 1902, p 36.
46. Barbara Ehrenreich and Deirdre English, 'The manufacture of housework', in *Capitalism and the family*, San Francisco, Agenda Publishing Co, 1976, pp 7-42.
    On the role of advertising and the media in persuading women to drop out of the employment market and strive for a domestic 'ideal' see the original classic: Betty Friedan, *The feminine mystique*, London, Penguin, 1968.
47. Barbara Ehrenreich and Deirdre English, *op cit*, pp 20 and ff.
48. John Bowlby, *Child care and the growth of love*, Harmondsworth, Penguin, 1965, pp 77-78.
49. Barry Hill, article in *Times Educational Supplement*, 14 January 1972.
50. Comer, *op cit*, p 143. For a discussion of Bowlby and 'maternal deprivation' theories, see pp 142-45, 149-64.
51. See eg World Health Organisation, *Deprivation of maternal care, op cit*.
52. See John Bowlby, *Child care, op cit*, Part I.
53. Ann Oakley, *Sex, gender, op cit*, p 183.
54. Ann Oakley, *Housewife, op cit*, p 210.
55. Margaret Mead, in WHO, *Deprivation of maternal care, op cit*, p 56.
56. US Department of Labour, *20 Facts on women workers*, n d.
57. Sullerot, *op cit*, p 119.
58. The news media are reporting instances of this phenomenon at the time of writing.
59. Sullerot, *op cit*, p 131.
60. See Barbara Rogers, *White wealth and black poverty*, Westport, Conn, Greenwood Press, 1976, Chapters I and II.
61. See Helen Hacker, 'Women as a Minority Group,' *Social Forces*, 1951, pp 60-69.

Chapter Two

Some analytical tools

'Status of women'

A common assumption among Western or Western-trained development planners is that the problems of women in developing countries are attributable to their 'status' in traditional society.[1] Hence the emphasis in the United Nations and similar organizations on 'raising the status of women' by involving them in development, often through special programs and projects.

The concept of 'the status of women' is one which conceals as much as it enlightens. It ignores the enormous variety of situations in which individual women may find themselves: according to position in the family, their own and their relatives' occupation, their income, among other such elements; the variations through time for each individual, relating to age, position in the household, health, and perhaps number of children; the fact that they have a varying relationship with individual men and boys according to the various male structures and life-cycles; and innumerable other factors. It is perhaps the most obvious indictment of the reliance on the concept of 'status of women' that there is virtually no parallel use of the term 'status of men' (although a few studies have been done in Africa on the 'status of men' in matrilineal societies). Finally, it seems clear that serious sociological studies on group status, which rely strongly on the subjective interpretation of people in the group concerned, are fundamentally different from the debate about 'status of women'. In this debate sweeping generalizations are made about half the people in a society based on statements either by outsiders or by the men in that society.

The analogy may be drawn with the situation of a black minority in a racist society. Paternalistic whites may spend hours arguing about 'the minority problem', 'the ghetto problem', perhaps 'the status of blacks', and debating what can best be done to overcome the problem of black people. Blacks, however, are more likely to see the problem in terms of whites ('honky', 'the man', 'whitey'), and the barriers they set up which limit the range of opportunities open to blacks. A disadvantaged group does not see itself as the problem; it sees rather the obstacles set up by others as the essence of the problem.

The concept of 'status' has often been used as an alternative to, or deliberate avoidance of, the concept of 'class'. However, the objections to a concept of women having a single, definable 'status' apply also to the idea of women as a class. In fact theorists of both tendencies have had difficulty in using their system to classify women.

The concept of 'status' is essentially a static rather than dynamic concept. The common formulation 'status quo' illustrates this, as does the fact that 'status' and 'static' derive from the same Latin root, *stare*, to stand. The connotations of this family of words may even be characterized as 'feminine', as that is characterized in European language and values. Women are objectified, or seen as passive objects which are acted upon by men; they do not act on their own environment or behave as agents of change in their relationships with men or with society in general, either as individuals or as a group or groups. It is highly questionable whether the men in most precolonial cultures allocated such qualities to women; and unlikely that women would see themselves in such a light. In fact, women in Western society, while consciously creating the illusion of their passivity, do not generally act as if they accept the ideology themselves. They are constantly initiating action, although often consciously disguising their own key roles in decision-making and implementation. They also take major responsibilities — for children, for example — and co-operate on a wide spectrum of activities with each other.

The concept of 'status', applied to a group of people — in this case, over half the population — by those outside the group, dehumanizes and objectifies people in that group, allowing no concept of their autonomy or independent action. Moreover, by applying this crude framework to which all members of the group are arbitrarily assigned, the concept of 'status' detracts from any serious consideration of the subtle elements that people do have in common: shared experiences, shared feelings, and similar kinds of action and reaction to a given environment.

The critique of the 'social order' approach in sociology, on which the concept of status is based, is of considerable interest here. As Alan Dawe summarizes it, 'the thesis that sociology is centrally concerned with the problem of social order has become one of the discipline's few orthodoxies.'[2] Sociology, shaped by the conservative reaction to change of the 19th century, is hierarchical with 'authority' among its most basic concepts; '. . . society must define the social meanings, relationships and actions of its members for them'.[3] Dennis Wrong has examined the history of sociology, seeing it as pursuing questions about the 'problem of order', assuming that there is a conflict between nature and society, these two being seen since Hobbes's time as co-existing and interacting opposites. Although modern sociology has forgotten the questions posed by Hobbes, it is still trying to formulate new and better answers to these old questions, its answers generally being divided between the notion of (a) 'internalization of social norms', and (b) the more common assumption that 'man (sic) is essentially motivated by the desire to achieve a positive image of self by winning acceptance or status in the eyes of others.'[4]

For women to internalize a super-ego representing social norms, the expectations and 'status' assigned to them by a male order, would be incompatible with mental health, or adequate functioning in the society even on its own terms. It could be argued that the predominance of women in mental hospitals and under treatment for depression is, in fact, related to their attempt to fulfil men's expectations about women, as the mother, sex-symbol, guarantor of family life or whatever – all based on a wide gulf between the male ideal and women's actual experience. George Brown and Tirril Harris have found a number of 'vulnerability factors' which predispose women to psychotic depression; they include having three children under 14 at home, and having no formal employment.[5] Both of these are part of the conventional 'housewife' role.

The second of the two concepts discussed by Wrong, that of the desire to win acceptance or 'status' from others, is also highly problematical. The word 'status' is used differently in this instance from that in 'status of women', it is an attribute to be won by positive – though conformist – action, rather than a category to which one is assigned by others. Here again, there is an assumption that approval, or 'status', is conferred by the dominant order, which is male; therefore for women to gain acceptance, they would have to conform to the male stereotype of what a woman is or should be. The concept allows no latitude for recognizing the alternative value-systems of women, whereby the approval of other women is based on quite different behaviour to that which gains approval from men. Nor does this formulation leave space for the phenomenon of women's conscious manipulation of male stereotypes in pursuit of their own distinct objectives, or the fact that women are usually very much aware of the split between male and female expectations about women. In other words, there is, in reality, no single scale of approval, or 'status', by which an individual woman can measure her own behaviour.

Control

An alternative approach to that of social order, and one which offers a more valuable framework for seeing how women function in different societies, is that based on the concept of control. In summarizing the difference between the two approaches, Dawe emphasizes the reliance placed by the first on external constraint; the key notion of the second is that of 'autonomous man' (sic).[6] The major concept here is that of control, which:

> '. . . adds the dimension of action to that of meaning; to control a situation is to impose meaning upon it by acting upon it. Secondly, it adds the dimension of interaction, or relationship between actors: to control a situation is to impose one's definition upon the other actors in that situation.'[7]

Here is a highly usable definition which accommodates the idea of different actors in society, for example women and men, acting according to different sets of meanings or values, which each has to modify to accommodate the sets of meanings held by the others. There may be illusions of control, or co-existing inconsistencies in sets of meanings (eg 'my wife doesn't work' and 'a woman's work is never done'). Where men are anxious to preserve at least the illusion of total control and conformity of all social phenomena to this set of meanings and myths, they will commonly explain away contradictory patterns of behaviour and the interests of women in terms of this supposed irrationality and illogicality. Great stress is laid on the influence of hormones, the menstrual cycle etc, as the basis for women failing to conform to male expectations.

In reference to Dawe's definition, however, a woman who attempts to act according to a male sterotype of women is controlled or objectified by it: ultimately an unhealthy approach both for her personally and for society as a whole. To function effectively in society, Dawe's approach seems to describe the essential precondition:

> '. . . actors defining their own situations and attempting to control them in terms of their definitions.'[8]

Returning to the overall social framework, Dawe suggests

that social systems should be seen as the result of a continuous process of interaction, which turns on the 'projects' and different capacities for control of the participants. The purposes which emerge as the result of interaction will then be the purposes of no single actor or group of actors, but will embody elements of compromise, and will involve unforeseen consequences.[9] Seen in this light, it would be of little value to assign fixed weights or ranking to different groups of people according to 'status' or rank in the pre-existing social hierarchy. The degree of control exercised by each individual or group is not quantifiable; social interaction is too complex, subtle and unpredictable for such a simplistic exercise.

Anthropology: a study of 'man and society'

Judgements by anthropologists and sociologists about the 'status of women' in other societies may tell us more about those who are making these judgements than about their subjects. It is rare to find Westerners attributing a higher 'status' to women of other cultures, even older women, than they do to women in their own. Friedrich Engels had few followers in his contrasting the 'barbarian lady', working hard and genuinely respected, to the bourgoise European who was 'estranged from all real work' and 'surrounded by false homage'.[10] The gap between perception and reality in the Western world on the whole question of gender-roles, outlined in Chapter 1 in terms of situational analysis, distorts Westerners' perceptions of these roles in other cultures. Although it might seem that attempts by anthropologists to step outside their own cultural preconceptions should enable them to take a fresh look at women and men in other cultures, there is little evidence that this has happened, with the exception of some outstanding work by some — not all — of the women in this field. The perceptions of Margaret Mead, to quote the most famous of these, have not been applied to anthropological studies as a whole. In fact, by focusing exclusively on women and children (an approach which arises from Western stereotypes about gender-roles in child-rearing) this work functions in a way seen also in the 'development' field — that is, it isolates women as a special case, fails to challenge the male bias in the rest of the anthropological literature, and provides an alibi for those involved. In other words, anthropology normally concerns itself with 'man and society'; this is supplemented by a few special studies on women in society.

Anthropology has always reflected Western viewpoints on

other societies; indeed, its efforts to understand other cultures make all the more striking the uncritical transposition of Western expectations onto these societies in terms of a man-based mythology. As Talal Asad stresses, anthropology is a product of the colonial encounter, becoming a flourishing academic profession towards the close of the colonial era; 'throughout this period its efforts were devoted to a description and analysis — carried out by Europeans, for a European audience — of non-European societies dominated by European power'.[11] The colonial power was, among other characteristics, run by men; and Asad's comment that 'there is a strange reluctance on the part of most professional anthropologists to consider seriously the power structure within which their discipline has taken shape'[12] could be taken to apply equally to their failure to recognize and eliminate the male bias of their discipline. Indirect rule, in particular, used local male hierarchies, and as Asad observes, 'at any rate the general drift of anthropological understanding did not constitute a basic challenge to the unequal world represented by the colonial system'.[13] Nadel tells us, without the least self-consciousness, 'In social anthropology as it is commonly understood, we extend our knowledge of man (sic) and society to "primitive" communities, "simpler peoples", or "pre-literate societies" . . .'.[14]

While the case against an anthropology of men cannot rest entirely on the linguistic anomaly of using this word in the two senses of 'person' and 'adult male', the influence of language in this case is certainly formidable.

The use of the word 'man' is ambiguous in modern English. Its derivation is of some interest. In Middle English, 'man' meant person, while woman and man were 'wif' and 'wer' respectively. The meaning of 'man' subsequently changed, with 'wer' becoming obsolete (persisting only in 'werwolf' — ie, man-wolf); its normal meaning in colloquial English is now 'adult human male, opposite to woman, boy, or both', as the Concise Oxford Dictionary formulates it. At the same time it retains its archaic meaning as 'person'; as is common with archaic formulations, this is in use mainly in formal or academic usage. Academic writers, in development studies as well as anthropology, can commonly be found using the word indiscriminately in both senses; sometimes (as in 'man the hunter') apparently as adult male, sometimes (as in 'primitive man') apparently as people in general. In virtually all academic works in development studies, or anthropology, there is a basic ambiguity about the terminology applied to people, which implies that males are the only people worthy of discussion, and that it is therefore unnecessary to be precise about whether one is discussing males or people in general. Occasionally the writers will acknowledge this problem by a brief reference to women as being outside the scope of their analysis. Thus Jackson, in his review of the main systems of stratification — power, status and class — cites the general problem of using such systems for categorizing women, unless they are among the unmarried minority.[15] As Wrong has pointed out, a common problem in sociology is to attempt to perfect answers without reference back to the original questions. If the questions originally posed are about men (with subsidiary questions about women), and if ambiguous and misleading assumptions behind the original questions are not re-examined, we might conclude with him:

> 'Forgetfulness of the questions that are the starting points of inquiry leads us to ignore the substantive assumptions "buried" in our concepts and commits us to a one-sided view of reality.'[16]

As long as the questions posed in Western writings about the Third World are in terms of 'men', the women will remain excluded from all the answers. This is a critical element in the process of discrimination in Third World development, since it enables the planners to see problems as entirely those of men; the solutions must therefore also be geared to the men. Without any individual intention or even thought about women, the overall effect is to exclude them absolutely from development planning.

Discrimination

In Parts Two and Three, on the nature of the development process and its impact on Third World women, we shall be considering those aspects or qualities of external influence and intervention which treat women as different from men, not because of their different 'traditional' activities and responsibilities, but because of a very specific Western (men's) model of what women in general should be, and what they should and should not do. The imposition of this model in terms of differential treatment of women and men respectively may be considered a process of discrimination. This has been described by Sheila Patterson as essentially a master of behaviour, rather than attitude:

> 'In the sociological context, this may be defined as the differential treatment of persons ascribed to particular categories . . . Discrimina-

tion then, is a way of behaving'.[17]

Thus it becomes unnecessary with a behaviour-orientated approach to consider the niceties of attitude and prejudice on the part of individual men, involved in development, towards women. This is consistent also with the most serious attempts to develop legal thinking on discrimination (on grounds of race and/or sex) which has recently evolved in United States jurisprudence: emphasis is placed on the discriminatory effect of an action on a particular group rather than on the intention behind it.[18]

Discrimination in the colonial heritage

The concentration upon men in the institution-building process under colonial rule is self-evident. Male hierarchies were used for direct or indirect forms of colonial rule, while female hierarchies atrophied or were actively suppressed, particularly by missionary organizations.[19] The new institutions, including the army, police, civil service, political parties, trade unions, churches, schools and universities, banks, local affiliates of multi-national companies and in fact all key development institutions are built up mainly or exclusively by male nationals. They are instructed by male expatriates in colonial administration, private enterprise, the colonial army or military advisers, development planning and administration, mission and state schools, and the rest. Much of the ideology of male supremacy is apparently passed on intact from the Western men to their local counterparts; the educated male élite — with some outstanding exceptions — are often much more hostile to women than are uneducated men.

It is worth noting that colonial administrators at all levels were men, even after women had begun to break into the professions and home civil service of the metropole in the 1920s. For the British colonies, for example, an Order in Council was passed in 1921 'to reserve to men any branch of, or posts in, the Civil Service in any of His Majesty's possessions overseas or in any foreign country'. This followed on the heels of the 1919 Sex Disqualification (Removal) Act, which ruled that women were to be allowed to 'assume or carry on any civil profession or vocation'. By the time women were officially admitted to the overseas Civil Service, in 1946, the Empire was coming to a rapid end, and the number of women serving, in very junior posts, was negligible.[20]

Colonial administration, then, was a man's world. The glamour of the Empire in fact helped to boost the hegemony of men back in the metropole; as Ruth Adam describes it:

'All the colour and romance of having an Empire — the beautiful uniforms — the great ships riding the storms — belonged to the men. The possession of all those faraway countries . . . depended on the men of the navy, the army and the colonial service. That made them an élite caste . . . They came and went between the homeland and that other world, the world beyond the seas, which was preached about, and sung and prayed about, but not seen . . .[21]

To an amazing extent this men's world was self-sufficient and segregated; its homosexual undertones have been called 'the best kept secret in British history'.[22] The successor caste to these Empire-builders — the development planners, administrators and 'experts' in the field — is on the surface much more integrated, with large numbers of women present; but it is in fact almost equally male-dominated at the professional levels. This is due to the extreme degree of domestication imposed on these men's wives, who are physically removed from employment opportunities of their own, together with the rigid career structure of the men's employment and the importance of wives' unpaid work in maintaining the family in an unfamiliar environment. Married women find it virtually impossible to join the system, particularly because of 'anti-nepotism' regulations and work permit restrictions, and also because of its career structure. At the same time, single women find themselves socially and professionally isolated and under considerably greater stress than their male colleagues.

The tendency for development institutions to remain virtually all-male at the professional level is of course reinforced by the fact that their counterparts in the national élite are men establishing their own family structure on the model of a Western-type nuclear family. The family is complete with a dependent wife who has even less prospect of an independent income to maintain herself at her husband's level of living than her Western counterpart.

Discrimination in the development process

The introduction of a cash economy and the pursuit of individual security through accumulation of material wealth (the opposite of traditional systems of using wealth as gifts to accumulate security through personal obligation) is also channelled through men. Cash crops and employment in plantations, mines, urban areas, and in many countries even paid domestic work, were overwhelmingly imposed on or offered to men, as was responsibility for serving in the colonial forces and payment of taxes. In the plantations of Asia and parts of Latin America, the whole family would be employed on the crop; however, it was the man who was regarded as the employee and the rest of the family had to work for him, either unpaid as in Brazil or at a small proportion of the man's wage as in the Asian tea estates.

With the trend towards Western-style ownership of land rather than customary and communal rights, it was women's rights to land that suffered most — both directly, because of colonial officials' failure to register women's assets or usufruct rights, and indirectly, because of their lack of access to cash for land purchase. As regards movable property it has been argued that, among horticulturalists, much of it is owned and inherited by whichever gender group has a use for it. Clothing, ornaments, shells, beads, cattle, pigs and other items used as special-purpose money, and cash itself where used, are controlled and inherited by whichever gender has the right to use them in organizing exchange or trade.[23] In other words, the means of production were controlled within the gender-group that used them.

The 'development' of a cash economy profoundly changes such a system, vesting ownership of the means of production in individuals regardless of whether they themselves work with them. A general pattern is for ownership of family assets to be vested in the 'head of the household', who is almost invariably classified as a man, even in cases where he is not physically present in the household. A combination of law and policy, directing the flow of cash to and from each household through its 'head', has had a profound effect in developing countries. Well-intentioned 'land reform' involves land being allocated to heads of household as outright owners; all other claims, especially women's, are suppressed. An outstanding example of the effectiveness of law and policy in this area is the suppression of bilateral or matrilineal systems of inheritance to a patrilineal system by instituting a new legal code and land tenure system.[24] The process is promoted by the concentration of cash and other forms of property in the hands of men.

Measures to 'protect', segregate and/or domesticate women often built on, or reinforced, local gender segregation and the division of labour. Colonial and Western influence makes sharper divisions between the sexes, particularly through education: scouting for boys, needlework for girls, special hospitals for women, separated public toilets. All help now to emphasize a person's sex, as Ester Boserup records.[25] Burmese women were in some ways much freer than is usual for European or many Asian women, and Margaret Mead has accused Western administrators of failing to take their true role into account. In Africa too, there have been innumerable complaints that missionaries of all kinds taught girls little more than domestic skills, if anything at all, and more or less encouraged a stay-at-home policy for urban women on moral grounds.[26] It was boys who were everywhere singled out for formal education, even in areas, such as most of West Africa, where Koranic schools had taught girls as well as boys.[27]

Excluded from education by missionary prejudice, women and girls have also been excluded from paid employment by entrepreneurs, although here the selection has been far less straightforward. In many areas where industries or mines were set up, women appeared to seek jobs, often outnumbering the men in the early years of industrialization, particularly in Latin America and the Caribbean and later in Africa.[28] A combination of the male institution of trade unions, working in close collaboration with men in government and the entrepreneurs themselves, contrived to drive many women out of industry by a combination of job reservation and government intervention (in Puerto Rico, Government grants were available on condition that two-thirds of the work force were men). In addition 'protective legislation' prohibited the employment of women on night shifts, from work with sophisticated machinery, and in situations deemed dangerous. In other words, Western gender stereotypes of women's 'frailty' were used to drive them out of many of the best-paid and least strenuous jobs in industry and mining, while hard manual labour continued in many cases to be allocated to 'low-class' women. Perhaps the most clear-cut correlation is between the mechanization of mining and industry and the dismissal of large numbers of women from the work-force. In Puerto Rico, Mexico, Argentina, Brazil and other Latin American countries, the initial high rates of participation by women in the industrial labour force were brought down to a small fraction within a couple of decades.[29] Where the 'protective' legislation might cause inconvenience due to the absence of men wanting the job in question — as in the case of nursing or night cleaning — it is not applied.

A similar, although less-well-documented process seems to be at work as regards independant entrepreneurs. As transportation networks expand women have become very active as traders, sometimes — as in Bolivia, much of West Africa, and elsewhere — operating over long distances and providing the major communications network for isolated rural villages. However, there are strong forces working to promote modern-sector trading, with enterprises owned by and employing mainly men, to take over from the markets where women often predominate. This frequently happens because politicians want to get into retail or wholesale trade themselves, as 'sleeping partners' or unidentified owners. Restrictions on informal trade help to dominate competition and street trading may then be made illegal, vendors harassed and fined, prices regulated by government, and licences withheld from most of the women traders. In some countries, notably Guinea, market women are a powerful organized force against this phenomenon; however, in most countries the informal trading sector is visibly losing out to the formal sector. There are some notable and well-advertised cases of wealthy women traders who have penetrated the formal sector, including import-export and small manufacturing businesses. Overall, however, they are an exception proving the rule (and moreover devote their profits to their sons' education for the modern sector). It has been suggested that a basic condition for the emergence of a substantial group of entrepreneurs, ie encouragement or at least tolerance by the government, is being largely withheld from women.[30]

An ideology of domesticity

At the same time as women are largely excluded from wage employment and other avenues into the cash economy, the ideology of their domestic destiny is strongly advocated, through the teaching of Western-type domestic skills and moral teachings about their place being in the home, etc. Development agencies tend to see women's participation exclusively in terms of 'home economics', based on the post-World War II ideology of 'maternal deprivation' and the forced return to the home of women working in industry and agriculture during the war. The doctrine of 'maternal deprivation', which determined the policies of the new United Nations development agencies, was attacked by Margaret Mead in 1954 as 'a new and subtle form of anti-feminism in which men — under the guise of exalting the importance of maternity — are tying women more tightly to their children . . .'.[31] More recently, advertising and consumerism have become strong reinforcements for the message of domestication and the 'consumption' role.

Probably even more important than the domestic or 'home

economics' ideology brought to women by development policies and programs, is the failure of development to bring significant benefits to the subsistence economy, of which women are
increasingly the guardians. Agricultural and rural development
programs are aimed almost entirely at men (often as 'head of
household'), a tendency reinforced by the trend for land holdings to be owned by men, which provides the criterion for
credit and other assistance. This kind of 'development' may, in
fact, intervene directly in women's subsistence activities in a
negative sense. It may increase their workload and in some cases
reduce their opportunities for earning cash income by diverting
land, labour and marketing outlets to cash crops, for which payment goes mainly to the men.

Dualism and dependence

The process by which women become involved in the domestic
subsistence sector, and men in the modern cash sector, is one
that closely approximates the dualistic model as described by
Clifford Geertz. He sees the dual economy as comprising a
capital-intensive modern sector and a labour-intensive traditional
one; the first develops rapidly and the second becomes rigorous-
ly stereotyped.[32] The former is also the export sector, the one
brought into being by colonialism and later neo-colonial relationships. As applied to Indonesia, the modern sector's expansion, stimulated by rising world commodity prices, was matched
by the contraction of the domestic sector of family-unit subsistence agriculture. Land and labour were taken from rice and
other staples and put into sugar, indigo and coffee, tobacco and
other commercial crops. When the export sector later contrac-
ted, responding to collapsing world markets, the second expanded again and a steadily growing peasant population tried to
compensate for the lost money income by intensified production of subsistence crops.[33] In other words, the domestic sector
remains a reserve economy for hard times and also subsidizes
the export sector through low rents and wages.

   The systematic bias of colonial and post-colonial administrations which reinforces dualism in a national or regional economy
can be illustrated by reference to Michael Lipton's formulation
of 'urban bias'. In 22 developing countries examined by Lipton,
barely 20 per cent of total investments in the period 1950-65
reached the agricultural sector, although it typically engaged 70
per cent of the population to produce 45 per cent of output.
This was the case even in 1960-65, before the 'green revolution',

investment in agriculture was, according to Lipton, associated with about three times as much extra output as investment elsewhere in the economy. Urban pressures have also caused the available agricultural investment to be misdirected, for example, to huge dams rather than higher-yielding minor irrigation works. Investment favours export and cash crops, and the rich urban food market, with little regard for calorie output per acre or per unit of investment. Above all, it has supported the large farm that supplies major urban outlets, although the small farm that feeds the rural poor makes more efficient use of land and other inputs, and hence produces a larger output.[34]

This concept of 'urban bias' to some extent parallels the 'male bias' in development, and provides a model for it. The urban and cash-farming economy is owned and operated by men, who in turn have more influence with policy-makers at all levels. The women in rich men's households are almost all in positions of extreme dependence. As a North India proverb, cited by O H K Spate puts it, 'To him that hath much, shall much be given'.[35] The domestic/subsistence economy receives a dwindling amount of resources and, as it falls behind, its political influence diminishes even more. Lipton's formulation of a consistent and cumulative bias in development (which he identifies as urban bias but which could similarly be seen as a form of male bias) is perhaps most useful in clarifying the increasing inequality inherent in the standard 'development' process, and another blow to the already discredited optimism of the 'trickle-down' enthusiasts.

The formulation leaves something to be desired, however, in that it portrays a process of increasing separation between more and less favoured sectors without much attention being paid to the powerful interactions between them, that help to sustain the momentum of inequality. The basic linkage here is supplied by the concept of dependence, or more exactly interdependence between the two sectors of a developing economy. The process has been described as 'internal colonialism',[36] a replication of the relationship between metropole and colony within the colony itself. It is marked by increasing dependence as investment and available skills are attracted to the centre (which is also the link with the metropole), with the dependent periphery receiving little or no investment. The main problem is the group of men making up the 'native bourgoisie',[37] which plays an important part in condemning the traditional sector as backward, unresponsive to incentives and fatalistic.

The concept of dualism as formulated by J H Boeke is perhaps too crude; it has been criticized particularly by Harold

Brookfield. He rejects as untrue Boeke's division of the Indonesian colonial economy into the Westernized sector, which was materialist, rational and individualist, and the Eastern element, characterized by self-employment, fatalism, unresponsiveness to variations in prices and wages, and an absence of profit-orientation.[38] This definition of two extremes is similar to the kind of opposing characteristics frequently attributed to the masculine and feminine spheres of activity.

Relations of exchange

A more useful analysis of the changing relations between women and men in the Third World is Ingrid Palmer's. She explains women's increasing dependency and relative poverty within the family as a reflection of the changing production and exchange relationships in the economy as a whole.[39]

Although in Western society 'the family' has an image of being a shelter and a private exception to public structures and relationships, such studies as are available indicate that the degree of power-sharing in private relationships within the Western nuclear family in fact reflect the resources — in terms of education and wages for example — available to each of the spouses from the society and economy at large.[40] Thus, Palmer explains the process of change in gender relationships in a developing economy, as connected to the relative access of each gender to material and other assets, such as wages or political influence. The 'generating core' of the connection is then the relations of exchange between women and men: exchange of economic powers, of familial powers, of other institutional and culturally determined powers, and of political powers. Each set of exchange relations governs the genders' link to variables found in the society and economy. For example, if education or family planning services are available, the set of relations of exchange between women and men will govern the access to and use made of them by each gender.[41] Palmer's prognosis is for increased inequality. Since women's work remains unpaid, and therefore uncapitalized and of low productivity, the inequality of exchange as regards the relative labour productivities of women and men will increase, tending towards a polarization, such as that observable between rich and poor countries, between urban and rural areas, and between different classes.[42]

The process as outlined above is self-sustaining in the sense that inequality of the relations of exchange is cumulative, as is urban bias in investment. New development institutions which,

in theory, could intervene counter to this trend are in fact rein-
forcing it. New institutions of monetization such as co-operatives and marketing agencies have virtually always given attention and prominence to the 'head of the household' as the selling
agent of the domestic unit, regardless of who is most closely
engaged in production. The reasons for this include, not only
bias as such, but also elements arising out of the increasingly
labour-intensive nature of female work: an example of this is
the fixed time-commitments that preclude them from participation at the hours and places arranged to fit in with men's work.
The problem is summarized by Palmer: 'The male takes on new
entrepreneurial roles as the custodian of family labour and
earnings, and his wife assumes some of the characteristics of a
rural proletariat'.[43]

It is important to remember that, in spite of women's increased dependence on men as controllers of family cash income,
and the diversion of resources (including women's labour) to
generation of that income, there is no reduction in women's
responsibility to deliver basic needs, such as food, to the
family.[44] This responsibility alone, comprising perhaps five to
six hours of non-field labour every day, enormously handicaps
women in their quest for independent sources of cash earnings,[45] just as Western women's domestic responsibilities handi-
cap them in the job market.

Although women are classified in terms of conventional
economics as 'dependants' of men, the real dependency relationship is two-way. As Brookfield describes it, the impression
is carefully fostered that the poor (or female) are dependent on
the rich (or male).[46] However, 'The desire of the wealthy [or
men] to retain and increase their affluence constantly augments
their own dependence — and hence also the aggressiveness with
which they seek to retain command over the system.'[47]

Once substantial inequality has been established in a society
undergoing 'development', the self-interest of those who benefit
most, and have the greatest power, will tend strongly to perpetuate and increase the divisions. Male bias has been built into
development institutions, processes and policies, and even if all
new programs were placed overnight on a foundation of equal
access for all, regardless of gender, the momentum of unequal
processes already in operation would remain very powerful.
Without wishing to accept the inevitability of inequality which
is implied in some dualistic analysis, we must realise that discriminatory processes, working against women in development,
are extremely strong. If we are to challenge them, we must first

try to understand how they function.

References

1. The relevant body in the United Nations is the UN Commission on the Status of Women, and its voluminous publications refer regularly to 'raising the status of women' as the major objective.
2. Alan Dawe, 'The two sociologies,' *The British Journal of Sociology*, Vol 21, 1970, p 207.
3. *Ibid*, p 208.
4. Dennis H Wrong, 'The oversocialized conception of man in modern sociology,' *American Sociological Review*, Vol 26, No 2, 1961, p 185.
5. George W Brown and Tirril Harris, *Social origins of depression: a study of psychiatric disorder in women*, London, Tavistock Press, 1978.
6. Dawe, *op cit*, p 214.
7. *Ibid*, p 213.
8. *Ibid*, p 212.
9. *Ibid*, p 213-14.
10. Friedrich Engels, *The origin of the family, private property, and the state*, edited by Eleanor Leacock, New York, International Publishers, 1972, pp 113-14.
11. Talal Asad, 'Introduction,' in Talal Asad (ed), *Anthropology and the colonial encounter*, London, Ithaca Press, 1973, pp 14-15.
12. *Ibid*.
13. *Ibid*, p 18.
14. S F Nadel, *The foundations of social anthropology*, London, Cohen and West, 1951, p 2.
15. J A Jackson, 'Social stratification: editorial introduction', in J A Jackson (ed), *Social stratification*, Cambridge, Cambridge University Press, 1968, p 3.
16. Wrong, *op cit*, p 183.
17. Sheila Patterson, *Dark strangers: a sociological study of absorption of a recent West Indian migrant group in Brixton*, London, Tavistock Publications, 1963, p 21.
18. Some recent publications include: Leo Kanowitz, *Sex roles in law and society: cases and materials*, Albuquerque, University of New Mexico Press, 1973. Kenneth M Davidson, Ruth Bader Ginsberg and Herma Hill Kay, *Sex-based discrimination*, St Paul, West Publishing Co, 1974, Supplement 1975. Barbara A Babcock, Ann E Freedman, Eleanor Holmes Norton and Susan C Rose, *Sex discrimination and the law: causes and remedies*, Boston, Little, Brown and Co, 1975.
19. See Rayna R Reiter (ed), *Toward an anthropology of women*, New York, Monthly Review Press, 1975; and Shirley Ardener (ed), *Perceiving women*, London, Malaby Press, 1975.
20. Ruth Adam, *A woman's place, 1910-1975*, London, Chatto and Windus, 1975, pp 99-100.
21. *Ibid*, p 9.
22. *Ibid*, p 12.
23. Ernestine Friedl, *Women and men: an anthropologist's view*, New York, Holt, Rinehart and Winston, Basic Anthropology Units, 1975, p 64.

24. Ros Morpeth and Patricia Langton, 'Contemporary matriarchies: women alone: independent or incomplete?' in Special Issue 'Women in anthropology,' *Cambridge Anthropology*, Vol 1, No 3, 1974, p 20.
25. Ester Boserup, *Woman's role in economic development*, London, George Allen and Unwin, p 219.
26. *Ibid*.
27. Prof Tremingham, article in *West Africa*, 1977, n d.
28. See Boserup, *op cit*; also June Nash, *Certain aspects of the integration of women in the development process: a point of view*, United Nations Document E/CONF. 66/BP/5, 1975.
29. A relatively large amount of literature is available on women in formal employment. See eg Boserup, *op cit*; and Nadia Yousef, *Women and work in developing societies*, Berkeley, Calif, Institute of International Studies, Population Monograph Series 15, 1974.
30. Harold Brookfield 1976, *Interdependent development*, London, Methuen, 1976, pp 86-90 and ff.
31. Margaret Mead, 'Some theoretical considerations on the problem of mother-child separation,' *American Journal of Orthopsychiatry*, 1954, pp 471-81.
32. Clifford Geertz, *Agricultural involution: the process of ecological change in Indonesia*, Berkeley, California University Press, 1963, p 53.
33. *Ibid*, pp 48-49.
34. Michael Lipton, 'Urban bias and food policy in poor countries,' *Food Policy*, Vol 1, No 1, 1975, pp 41-52.
35. O H K Spate, Introduction to T Scarlett Epstein, *South India: yesterday, today and tomorrow: Mysore villages revisited*, London, Macmillan, 1973, p xiv.
36. See James D Cockroft, André Gunder Frank and Dale C Johnson, *Dependence and underdevelopment: Latin America's political economy*, Garden City, NY, Doubleday, Anchor Books, 1972.
37. Brookfield, *Interdependent development*, pp 163-65.
38. *Ibid*, pp 54-55.
39. Ingrid Palmer, *Monitoring Women's Conditions*, unpublished draft for UNRISD, n d.
40. Lee Comer, *Wedlocked women*, Leeds, Feminist Books, 1974, pp 66-67.
41. Palmer, *Monitoring women's conditions*, pp 50-51.
42. *Ibid*, p 43.
43. *Ibid*, p 42.
44. *Ibid*, p 23.
45. *Ibid*, pp 41-42.
46. Brookfield, *Interdependent development*, p 189; see also Chapter 7.
47. *Ibid*, pp 202-3.

Part Two:
Discrimination in
Development Planning

Introduction to Part Two

In Part One, a range of different ideas and ideologies were discussed which relate to the Western male tradition of 'a woman's place'. Among other things the central concept of discrimination was introduced, with emphasis not on the motives involved but on the discriminatory effect. In the world of development researchers and planners, women are so thoroughly excluded from the analysis — except in the fringe area of 'social development' — that the planners can convincingly express surprise that they should be accused of practising discrimination against them. Part Two examines some of the mechanisms by which development organizations and individual planners work a discriminatory system without awareness of what is happening.

Chapter 3 examines discrimination inside the international development agencies, with examples mainly from the United Nations Development Program (UNDP) and the UN Food and Agriculture Organization (FAO), which effectively relegates the few women in professional posts to very low levels while most of the women involved serve the men as secretaries and wives. This is followed by a review, in Chapter 4, of the methods for collecting and classifying data about Third World economies and people in the conventions of statisticians and development researchers. Finally, Chapter 5 discusses the problems associated with the existence of a marginal women's sector in development institutions, channelling many of the best Third World women into Western-defined subjects such as 'domestic science' or 'home economics'. The problems of using special women's projects in the attempt to promote integration of development projects and programs are also outlined, together with a discussion of the 'population' projects aimed at women.

Chapter Three

Inside the international agencies

Discrimination in development agencies

Colonial administration, as we have already noted in Chapter 2, was essentially a men's club. Even at a time when women were entering a wide range of professions and particularly government jobs, which in Britain followed the 1919 Sex Disqualification (Removal) Act, they were officially barred from any part of the overseas Civil Service. The pattern today is very similar; of all the professional employment opportunities available to women, perhaps the greatest difficulties and the most severe forms of discrimination are to be found in the international sphere, including development planning for the Third World.

The 'pattern and practice' of sex discrimination at United Nations headquarters in New York is cited as a classic case of infringement of United States anti-discrimination law, and it is protected from prosecution only by its immunity from American law.[1] The nature of this pattern is by no means accidental, either, as has been shown by the violent reaction of personnel and other senior officials to challenges from women's groups which have been formed in many of the UN bodies. A case in point is the United Nations itself which, apart from its well-known political functions, is responsible for the UN Economic Commissions, for much of the technical assistance provided through the UN network, and for many other areas of development policy.

In 1975, the UN's official International Women's Year, a small group of women at the United Nations in New York drew up a questionnaire on attitudes and perceptions about discrimination, which was distributed to a random sample of employees, both men and women. The response of the Personnel Division was immediate: returned questionnaires were seized as they reached the internal mail system and an immediate witch-hunt was started through the plain-clothes security guards to identify the members of the group, with threats of dismissal and other forms of disciplinary action. Particular anxiety was expressed by senior officials about the questions dealing with sexual harassment of female employees by men, both officials and delegates to the United Nations, and the fact that some of the returned questionnaires had named some of the more notorious offenders. The administration failed to discover the membership of the group, and suffered severe embarrassment as the incident was published in the American and other press.[2] A senior official later described discrimination against women as a 'very dangerous' issue for the United Nations.[3]

A useful body of information now exists on discrimination against women in the United Nations system. Data from the United Nations itself, the UN Development Program and the UN Food and Agriculture Organization (FAO) are of relevance, particularly since all three have substantial responsibility for rural development planning on an international scale, and are also responsible for standardization of procedures at the national level.

In the FAO, where the first survey was carried out, replies from over 800 women indicated that 56 per cent of them felt they were treated as inferiors; some three-fifths reported that they were required to provide additional services for their male superiors which were outside their official job description; over 50 per cent considered that they suffered discrimination in terms of promotion, that they had to work harder than men, and that they had lower titles and grades than men doing the same work. There was felt to be explicit discrimination against married women.[4]

A subsequent and more comprehensive survey of the United Nations in New York produced about 900 replies (an excellent response rate for this kind of survey — about 25 per cent). Fifty-seven per cent of the women who responded considered that their career prospects had been influenced by their gender, as opposed to only 28 per cent of the men. Of the female professionals, 86 per cent felt handicapped by being women. Areas of discrimination included recruitment, promotion, work assignments, overseas travel, the imposition of additional work outside the job description, and sexual harassment. The highest proportion of complaints about discrimination was recorded in the Office of Personnel Services. Generally speaking, it was striking that in answers to many of the questions, women and men had completely different perceptions of the problem; for example, 55 per cent of the men considered that the position of women in the UN during the 1970s had improved, as opposed to only 23 per cent of the women.[5] Given the enormous preponderance of men in decision-making positions, this suggests that there is a systematic failure in this kind of organization to perceive that women face discrimination.

A United Nations Development Program (UNDP) report mentions a problem which is common to all the United Nations development family: the extreme rarity of women from the Third World holding supervisory or managerial jobs.[6] This is a factor which can be readily observed in any of the agencies, at the field level even more than at headquarters; and it is this absence of women from the countries most directly concerned in the development process that is perhaps the key to understanding the failure of development planning organizations to recognize the central role of rural women in the Third World as a whole. UNDP is now making efforts to recruit more women, but it will be many years before these women move into decision-making posts.

A model of dependency for the planners

One of the most consistent responses of male planners to the introduction of discussion about women in development, is to base their argument against change on the domestic model familiar to them: '*my* wife doesn't work', 'I get on very well with my daughters', 'my mother always said a woman's place is in the home', and other variations on the theme that all women should follow the model of feminine deportment which they consider correct. This is not necessarily a true reflection of what the wives', daughters' or mothers' lives are really like, and the planners tend to become anxious if the reference to their wives is followed up by the suggestion that the wives participate in the debate.[7]

There is a further, often heard rationalization: that local governments or 'tradition' would not tolerate any attempt to involve women in development. This is used with surprising frequency by Resident Representatives as the excuse for refusing to accept a woman on their staff, even when nominated by headquarters for a post. It is flatly stated that the government concerned would not approve the involvement of women and therefore none should be nominated; and it is not generally thought necessary to attempt to verify such statements. Recently, for example, a regional representative for the United Nations Childrens' Fund (UNICEF) circulated a letter inviting applications for a job in East Africa, which stated that only men need apply, as governments such as those of Burundi and Madagascar, supposedly, would not be able to work with a woman.[8] No evidence for this assertion was produced and in fact the countries concerned are those with a relatively high concentration of professional women, working for bilateral and international development agencies.

In UNDP, although the policy of the Division of Personnel is that 'No staff member should be indulged in a preference for a male or female supervisor', apparently there has been no such policy in terms of preferences for male or female subordinates. The 1977 Report observed that, even while routinely posting women to some of the more difficult duty stations, the Division of Personnel 'are simultaneously refusing to consider other postings for which professional women apply, on the grounds that the post is "inappropriate for a woman" '.[9] Similarly, 'the preference of some Resident Representatives for all-male staff in positions other than local staff is another contributing factor to the present staffing profile and the recruitment statistics.'[10] At FAO, rejection of women as professional staff members or as experts is considered quite normal. One case cited to me was that of the FAO representative in Turkey, who refused an expert on the grounds that a woman was unsuitable even though the government had already accepted her.[11]

Claims that certain governments — particularly in the Middle East and sometimes those in Latin America — would refuse to deal with professional women are in fact based largely on Western stereotypes about the countries concerned. Sudan is a typical case, a country on the list of many governments and international agencies as 'unsuitable' for female representatives who, it is claimed, would not receive the respect due to the organization they represent. When asked about this, Sudanese Government officials expressed astonishment that any such assumptions had been seriously used, and felt that this was a gross insult to them. They said that, if anything, women from other governments or intergovernmental bodies would be given more respect and attention than men.[12] The author has found this to be generally true in contacts with Arab men at the United Nations, although it is usually the Arab countries which are most consistently cited as refusing to work with women.

The discrimination process is surprisingly efficient at eliminating women; there are virtually no women working as experts for the Technical Assistance Recruitment Service (TARS) run by the United Nations for the UN family as a whole. All experts have to be approved by a series of people at headquarters and in the field. The virtual elimination of women extends even to subjects where they predominate in terms of numbers and experience, such as social work or tourism.[13] Thus an important source of technical expertise is being arbitrarily rejected.

'But that's not the real issue: it's Third World women we should be concerned about.'

The above statement is heard with surprising frequency as a reason for refusing to work on cases of discrimination in planning organizations, no matter how well-documented. It was an important part of the reasoning behind the refusal by the Secretariat of International Women's Year, and the ensuing Decade, to intervene in any internal discrimination cases, or even to take an interest in the composition of UN staffs at all. In fact, the professed concern for women who are so far removed from United Nations headquarters that there was no possibility of their participating in the debate has yielded very disappointing results. One of the most noticeable effects of 1975's International Women's Year was a backlash among men in the UN and its agencies who expressed themselves bored and alienated at the mere mention of the word 'women'.[14] This was applied equally to Third World women and to female planners. Since no significant advances were made in that year in terms of women's participation in the planning agencies, the hostility of the male planners has remained a significant barrier to the involvement of Third World women in development.

Male planners and male bias in planning

Women working as professionals in planning organizations, regardless of their degree of identification with the women's movement, can be observed to be disinclined to describe the world in terms of 'men' as a synonym for people. They are also less likely — although it does sometimes happen — to express boredom and hostility with the question of women in development. The problem of the few relatively senior women who have adapted to a male system and ideology by rejecting any identification with women as a group has been, and to some extent remains, a problem.[15] However, a number of these senior women have been observed to modify their position quite substantially, as a result of the publicity and action over discrimination in the agencies against women.

Even more important, however, is an ending of the sense of isolation for women in middle management, which has made them so defensive about their work. The prospect, if not yet the reality, of more women joining their organization at the middle level, or being promoted through the ranks, has a very important effect in headquarters positions. In general, the largest agencies feel this effect first, since a small proportion of women produces a total number which gives what might be termed the 'critical mass' necessary to counteract the sense of isolation. This factor is so important that Rosabeth Moss Kanter, in an important paper on promoting equal opportunities for women in public service systems, advocates that managements refrain from distributing additional women, recruited under affirmative action programs, around the entire organization in order to provide a token woman in each office. Tokenism, she observes, handicaps any member of a racial, ethnic or gender minority who is working nearly alone among members of another social category.[16] Such people are conspicuous, particularly in their mistakes; and 'they face pressure to side with the majority group against their own kind, as a price of membership'. In addition, they may secure acceptance by adopting a stereotyped role attached to their group: for example, as mothers, sex objects, daughters or younger sisters, or de-sexualized militants, thus limiting themselves to only a small part of their total personalities, and isolating themselves from the informal contacts which are essential to the smooth functioning of an individual worker or an organization. They are also subjected to extra stress, such as the well-known habit of working twice as hard as men in order to 'prove' that they are as competent. Kanter recommends the clustering, rather than spreading, of women brought into previously all-male offices.

This recommendation highlights what is perhaps the major issue in linking current employment patterns for professional women with the failure to integrate women in grass-roots development: many field offices of development organizations have no women at all, as planners, as project managers or as 'experts'; of the rest, the overwhelming majority have only one each in one or more categories. The result is the most extreme form of isolation and what Kanter sees as tokenism. One manifestation of this is that a woman working in a field office is almost inevitably assigned that low-priority set of projects or activities categorized as suitable for women (a categorization which is examined in more detail in Chapter 5). As far as the important work is concerned, this is effectively handled almost exclusively by men. The men's perception of the people they are supposedly helping are therefore of central importance in moulding the patterns of intervention as they affect women. As Conrad

Reinig has observed, the staff and organization of a development project or program should be as much a subject of enquiry
as the 'people being developed'.[17] Robert Chambers goes even
further:

> '. . . in trying to understand projects and to derive practical lessons
> from them the staff and their organization are, if anything, more
> important than the people they affect. It is the staff who decide
> policy and execute it. It is they who perceive or fail to perceive the
> details of the situation in which a scheme is launched. It is they, and
> not the people being developed, who hold the initiative, especially
> in the early stages of a project. If staff and organization are ex-
> amined . . . as a primary focus, then more practical lessons may
> emerge. Developers may be able to learn something about their own
> behaviour, about the problems and conflicts they are likely to face
> and about needs that have to be anticipated.'[18]

All the factors have to be taken into account, he adds; 'The
roles, attitudes and behaviour of both staff and settlers are thus
included . . .'.[19]

The men in the field

The commonest reaction among male planners in the field who
are asked questions about the women is: 'I never thought about
that'. In some situations, however, the attitude is somewhat
modified by experience: occasionally, project staff and field
representatives with long exposure to the area will tend to have
a relatively accurate perception about local women, certainly
more than their counterparts in headquarters. In Zambia, the
senior UNDP and FAO representatives, all with long experience
in the country, vied with each other in describing to me the preponderance of agricultural work done by women; their success
as growers on the scheme, once having established the right to
participate in it; and the outstanding record of their female
agricultural officer. The men, they agreed, did little or no work
but spent the time drinking and talking politics.[20] More commonly, the difference between field and headquarters staff is
that the former will be more amenable to answering questions
on the subject. This suggests that returning to the issue on a
number of subsequent occasions could be very productive. For
example, following a question about the contribution of women
to particular crops, it would be more likely that a question
designed to produce this information would be included in subsequent surveys. Querying the fact that the incidence of a
disease is expressed only in terms of the proportion of men

afflicted, which had previously never been noticed, could make it embarrassing for planners to continue the practice. A few examples of conversations with male planners in the field may be used to illustrate the process.[21]

An FAO country representative:

> 'I've just been filling in a questionnaire from headquarters about women. But you know there's hardly anything to say, because we don't have the sort of projects that would involve them. We have nothing against them, in fact we'd like to have more for them, but you see all our projects here are concerned with cattle, and it just so happens that women have very few cattle. Of course, we get criticized because cattle are owned by the richer people.'
>
> 'Is it perhaps more than just a coincidence that all the money is going into cattle and almost nothing for crops, when cattle are men's responsibility and crops are women's?'
>
> 'I never really saw it like that. But I suppose there is a connection.'

Senior officials of a World Bank project:

> 'Meet Barbara Rogers, she's visiting this project and wants to know what we're doing for women. I warn you though, she's a feminist.'
> Embarrassed silence.
> 'Well, actually I don't think there's anything of much interest to you here. Perhaps UNICEF can show you something. We're a huge program, millions of dollars, a consortium of agencies, got a job to do, and we haven't got any time for special projects."

A voluntary agency's regional representative:

> 'We have this gap between our publicity people and the field. For example, they keep saying we dig wells here to avert another drought, and we stopped doing that some time ago.'
>
> 'Don't you believe in wells any more?'
>
> 'Well no, we've had very negative experience with this well-digging business. Everybody thinks they're so important, but we've done everything we can think of and we can't get the community to take responsibility. I've been to villages, collected all the men together, talked with them for hours. Then I come back with the equipment. We dig down until midday, and there's some water at the bottom of the hole. I can't get them to understand that they've got to keep on digging, otherwise the well will dry up as soon as the dry season starts. They just refuse to dig any deeper, and of course their wells dry up and they come and ask for more help. I'm fed up with wells.'
>
> 'But why are you only talking to the men in the first place? They're not the ones who go to the well every day, year in and year out, and who actually observe the level going up and down according to the season; the women do.'
> Pause.
> 'I didn't think of that then. That was before I started to become enlightened about women.'

Onchocerciasis Control Program (World Health Organization and others):

> 'Why do all your statistics on the incidence of river-blindness only refer to men? Don't women get it? Or aren't you concerned about whether they do or not?' (A question repeated several times.)
>   Three different replies:
> 'There isn't any difference. Women and men get the disease at exactly the same rate, as far as I know.'
> 'In our village visits we find it's really a men's disease. Sure, we'll find three or four blind women in each place, but it's usually trachoma, not oncho. We might treat them, but only to get the villages' co-operation for our oncho program.'
> 'There's no common pattern. My sociological data show that it depends entirely on the seasonal work patterns, who works close to the river for the longest periods. In some villages more women have oncho than men. There are so many factors: for example, I find that if there's a small well in a village, and the women don't have to go to the river for water, they're less exposed to the fly.'

(I asked several of the officials if there were any plans to provide wells for badly affected villages. The answer was an adamant 'No'. This was a serious eradication project; they were spraying the blackfly, and that was that. In fact, the eradication program is a great deal more precarious than this would suggest.)[22]

Finally, a more critical official:

> 'The documents only mention men having the disease? I don't believe it. Let me look it up . . . Well, how amazing! You're right! D'you know, I've worked on this for years, read that document dozens of times, and never noticed that it's only about the men. And now I come to think of it, our film only shows pictures of men. But surely the women get it too. I wonder why it was done like that . . .'

Director of a co-operative employment project:

> 'I've tried to introduce vegetable production here, too. I got this man to try it, spent hours with him suggesting new techniques, working with him, persuading him to experiment with different methods. Yes, it looks very impressive, and he gets a good income from it. But I'm very disappointed that it hasn't been taken up by anyone else in the village, which was my original idea.'
>   'Didn't any of the women show any interest?'
> A long silence. He sucked his pipe. Finally, very slowly, he replied. 'Perhaps if I'd tried it with the women in the first place the idea would have got a lot further by now. You know, I just never thought of that, can't think why. Now it's too late.'

These conversations, and several others similar to them, indicate that male planners do not recognize the participation of women in important work, even when it is in front of their noses; that when questions were asked, their responses range from denial to surprise that they had never been made conscious of the problem; and that if a series of questions could be put at various stages, or if there were more discussion of women in general, their level of awareness could be improved. The result of this would be reflected in the way they conducted surveys, wrote up projects and program documents, and implemented them.

The problems of educated Third World women 'locally employed' in their own countries by the United Nations, although not so far analyzed in detail, are perhaps the most serious of the whole discrimination issue: these women are handicapped both by being 'locally employed' and by being female. The UNDP's Working Group expressed its 'concern for the conditions which were brought to its attention' and emphasized that for this group 'the issues are far too serious and conditions far too widespread for the Working Group to do justice to them in the week or so between when letters arrived from the field and the date when the report had to be finalized.'[23] It is this problem, together with the lack of Third World women on the international staffs, which is the most serious gap in any attempts by the development organizations to reach women at village level.

References

1. A conclusion reached by Professor Ruth Ginsburg at the Law School of Columbia University, New York, and used in her lectures on anti-discrimination law. See Kenneth M Davidson, Ruth Bader Ginsburg and Herma Hill Kay, *Sex-based discrimination*, St. Paul, West Publishing Co, 1975.
2. See e g 'UN brass confiscates a sex quizz', *New York Post*, 6 October 1975; 'Women employees at United Nations set "solidarity rally" ', *Wall Street Journal*, 4 December 1975; 'Women on staff at UN fight "male chauvinism" ', *New York Times*, 10 December 1975.
3. From a conversation with the head of one of the UN bodies.
4. *Summary of survey results*, FAO Women's Group, Rome 1975.
5. Ad Hoc Group on Equal Rights for Women, *Survey of staff attitudes on sex discrimination in the United Nations Secretariat*, 1976. 1: Analysis and recommendations for action. 2: Statistical tables.
6. UNDP Staff Association, *Report of the working group of the Staff Council on the status of women in UNDP*, Staff Council Circular UNDP/SCC/HQTRS./26 and UNDP/SCC/FIELD/15, 28 November 1977, p 13.
7. This observation has been made by a number of professional women who are trying to work within international development agencies to

raise the question of women in development, Mallica Vajrathon wrote about it as a phenomenon in meetings in the field in *UNDP News*, March/April 1975. Fran Hosken, who worked as a consultant to the World Bank on site and service schemes, has told me that she found references to wives the almost universal response to her attempts to discuss women's role in the settlement projects.

8. I was shown this letter by a man employed by UNICEF; when I said that I intended to challenge this piece of discrimination, he changed his mind about giving me a copy. The immediate effect of women challenging blatant discrimination is often that the men start to conceal its operation as a means of ensuring its continuation.

9. UNDP, *Report of the Special Committee*, p 24.

10. *Ibid*.

11. Information supplied by members of the Women's Group in FAO.

12. Conversations with officials of the Ministry of Foreign Affairs, Khartoum, December 1973.

13. Information given at a meeting between TARS representatives and the Ad Hoc Group at the United Nations, New York, 1976.

14. This theme is elaborated in Barbara Rogers, 'International men's year,' *UNDP News*, March/April/May 1976, pp 7-10.

15. Oreo is the name of a biscuit (cookie) widely sold in the United States which is black on the outside and white on the inside.

16. Rosabeth Moss Kanter, 'Changing organizational constraints: toward promoting equal opportunity and treatment for women in public service systems', *Public Administration and Finance Newsletter* No 59, September 1976 — June 1977, p 20.

17. Conrad Reinig, *The Zande Scheme: an anthropological case study of economic development in Africa*, Evanston, Northwestern University Press, 1966, p 214.

18. Robert Chambers, *Settlement schemes in tropical Africa: a study of organizations and development*, London, Routledge and Kegan Paul, 1969, pp 8-9.

19. *Ibid*, p 9.

20. Conversations with the author in Lusaka and Kalulushi, Zambia, September 1977.

21. These conversations all took place in the course of visits to UNDP and other field offices and projects in Africa, August — October 1977.

22. See Barbara Rogers, 'The river-blindness gamble', *West Africa*, No 3150, 21 November 1977, pp 2348-49.

23. UNDP, *Report of the Special Committee*, p 28.

Chapter Four

The treatment of women in quantitative techniques

Introduction

Planners do not deal with individuals; rarely do they have direct contact with peasant communities or their representatives. The data-base for planning is the statistics, surveys, censuses, and their derivatives such as 'manpower plans' and cost-benefit analyses. As development planning and administration in the Third World is subjected to a continual search for improvement and refinement of analysis, particularly with the use of quantitative techniques, the dependence on formalized data is increasing. It is therefore important, in seeking to evaluate the impact of development planning on Third World Women, to examine how these data reflect their economic importance.

National accounting procedures

One of the earliest efforts of the United Nations in the economic field, which has been outlined by Brookfield as a key to international 'development' initiatives, was a drive to create a vast improvement in national statistical services in Third World countries, and to back this with a system of international statistical collection, standardization and reporting.[1] The drive was initially sponsored by John Maynard Keynes, and his continuing influence led to the trend to aggregate national income through estimates of Gross National Product (GNP) and other measures, at the expense of income distribution within the country.[2] National income data became available for a large number of countries and territories by the late 1950s. The World Bank, which in 1956 shifted to lending mainly to Third World countries after its initial post-war emphasis on Europe and South Africa, led the way to the unfolding of a series of development plans, based on the new data; and this obviously gave a boost to the formalization and standardization of national accounting procedures.

The United Nations statistical agencies accomplished one of their major successes in a drive to achieve a vast improvement in worldwide statistical coverage for the census years 1960 and 1961. By 1970 few 'market economies' remained for which at least estimates were not available.[3] These indices have become of central importance for evaluating and planning the economic performance of countries, and have passed almost into everyday use, as well as providing the data base for national development planning and the allocation of resources in general.

Considerable vested interests are now involved in preserving the standardized system of GNP evaluation, since membership of certain categories of low-GNP countries such as 'least-developed' or 'most seriously affected' (by the rise in oil prices) is a guarantee of access to considerable international and bilateral aid on softer terms than generally applicable. The discouraging GNP data which led to the efforts to promote 'growth' in Third World countries in the First Development Decade, are now very hard to challenge despite the growing recognition that they reflect, more than anything, the wholesale exclusion of informal and non-cash output in Third World countries: the sectors in which women are particularly concentrated.[4]

Despite the growing criticism, in recent years, of the use of the conventional national accounts for measuring the levels of living or welfare of the population (or even its true output), which is acknowledged by the United Nations, no real change is likely. Perhaps the most fundamental objection, that non-monetary economic activity should be included in the estimates, was brushed aside as a 'welfare' issue by the Expert Group on Welfare-orientated Supplements to the National Accounts and Balances and Other Measures of Levels of Living, which met in New York in 1976.[5] The UN Statistical Commission also, while recognizing the importance of the problem for purposes of 'welfare' measurement, dismissed the importance of informal and non-monetary production. According to the Report of the Expert Group:

> '. . . the Commission stressed that work on welfare measurement should not be allowed to disturb the SNA and MPS. Measures of welfare should be developed as supplements to the national accounts and balances, not as revisions to them.'[6]

By explaining away women's work as if it were a measure of welfare rather than an element in production, the international statistical authorities — and this is one area where the United Nations has considerable influence — are making it clear that there will be no interference with the Keynesian approach to output. There seems to be a strong conservatism attached to the whole area which is even more marked in the international agencies, and in Third World countries, than it is in some Western countries where the problem of non-monetary production is receiving considerable attention at the highest level. The value of women's domestic work in the United States, for example, has been the subject of a number of serious studies by the US Government and by economists of the standing of John Kenneth Galbraith.[7]

These critiques point out that a major part of the economic activity omitted from GNP calculation is women's work: in Western countries, mainly unpaid domestic service, and in Third World countries mainly subsistence food production (and in the towns, a wide variety of informal activities). Galbraith suggests that the domestic services of housewives in the United States are worth roughly a quarter of its GNP. Even more important, and concealed, is that 'the servant role of women is critical for the expansion of consumption in the modern economy' since otherwise household consumption would be severely limited by the time available to employ people to manage such consumption — to select, transport, prepare, repair, maintain, clean, service, store, protect, and otherwise perform the tasks that are associated with the consumption of goods.[8] The conventional family lifestyle would become untenable.[9]

A similar situation exists in the Third World, particularly in rural areas. Although the actual work performed is quite different, and cannot be accurately described as 'domestic', it remains unpaid and therefore unvalued by statisticians and planners of the modern, commercial economy, though it is, in fact, essential to the very existence of that economy. In this case the contribution is mainly in the production of basic goods rather than consumption of marketed products. The less 'developed' the national economy, the more crucial the non-market production in which women play such an important part. Such an economy of course appears as poverty-stricken in the extreme, since the relative importance of the market economy is small, and that is practically all that is counted in the GNP. International aid is then concentrated on this 'least developed' country to build up the cash sector at the expense of the unvalued subsistence sector, and the system of standardizing and internationalizing 'development' in the Third World is given a considerable boost.

The GNP and its variants are, to a large extent, a statistical illusion. It refers ostensibly to production; and increasingly it is being used, despite misgivings, as an indicator of national welfare. However, as Carolyn Shaw Bell has summarized it, 'although GNP refers to what is *produced*, it is usually figured by what is *bought*'.[10] By definition, then, if women's productive work is not paid, it has no place as real production. While this is a problem which has been discussed mainly in terms of domestic work performed by women in Western countries, it is even more serious for those in the Third World. An International Labour Organization (ILO) employment mission in Kenya summarized the problem as follows:

> '. . . the sharp distinction between time spent on "economic" activities and on work for the family (fetching water, preparing food, teaching children and dealing with their ailments and so on) is ultimately arbitrary; many of the services just listed do more to increase family welfare than services formally counted as economic . . . Although these problems have long been recognized in the literature on national accounts, they require a new and more pervasive significance when one begins to concentrate on the poor in predominantly rural countries.'[11]

The labour force as 'manpower'

The same, it would appear, applies to labour statistics. Women who are not paid for their work are defined as non-productive, and in this case they are seen quite differently from men in a similar situation. Labour statistics characteristically portray a large proportion of adult men in any given country as 'participating' in the labour force (a concept which is much wider than that of wage-employment), while a very small proportion of adult women are seen as 'participating'. Gustavo Pérez-Ramirez, Chief of the Operations Section of the UN Population Division, considers that defective coverage of women in many national data collections is especially marked in the case of the labour force.[12] He points out that only 28 per cent of all women and girls are considered 'economically active', as members of the labour force, as compared with over half the men and boys.[13] If the statistics were to be believed, it is the Third World countries where women are least 'economically active': for example, only four per cent in North Africa and 11 per cent in tropical South America.[14]

An attempt was made in the ILO Employment Mission to Kenya to evaluate the usefulness of labour force statistics as applied to women's work. It concluded that in a country like Kenya, where the overwhelming proportion of the population lives in rural areas with very low incomes, preoccupation with conventional definitions of the 'labour force' becomes pointless:

'In official Kenyan statistics, it was assumed that 45 per cent of women "participated" in the labour force. The reality is that throughout the rural areas most women are working in the field — usually for much longer hours than the men.'[15]

The problem of defining labour force participation is described as particularly difficult in the case of women in the rural areas, where the demarcation line between 'economic' and 'non-economic' activities 'is statistically arbitrary and for purposes of indicating living standards, meaningless.'[16]

Meaningless though such distinction may be, they in fact lie at the heart of planners' categorization of the labour force which treats women as marginal workers, if indeed they exist at all. Women in agriculture are commonly placed in the vague category of 'unpaid family labour', which also includes children who may work widely varying amounts of time. Thus we may be told, in what purports to be a serious study of workers in the Philippines, that agricultural workers included 3,075 million 'self-employed workers' and 2,492 million 'unpaid family workers'.[17] The former category probably includes mostly men smallholders; the latter is misleading since it covers women and children together, and may well be a major under-estimate of the numbers of both categories.

Analogous to the use of the 'family labour' category is the habit, among most writers about Third World agriculture, of reserving such terms as 'farmer', 'peasant' or more fanciful terms such as 'husbandman' for men only. Women, if mentioned at all, are 'farmer's wives'.[18] The habit is unaffected by the fact that women may do much, or even most of the farming, and may show more characteristics of a peasant than the more itinerant men. Jurion and Henry provide a classic in this kind of double-think in their tome on African agriculture. They attribute the invention of agriculture in Africa to women, and stress the fact that women do all the agricultural work, with help only from their children; 'in undeveloped Africa, therefore, woman is still the true farmer . . .'. This does not prevent them from referring to the men thereafter as farmers and the women as 'the farmer's wife'.[19]

'Head of the household'

The relegation of women to the miscellaneous category of 'family labour' is consistent with another of the fundamental concepts transferred from Western society to planning for the Third World: the concept of the 'household' as the basic unit of society, represented or managed by a 'head' who would normally be a man.

Galbraith has characterized the Western concept of a 'household' in non-classical models of an economic society as an extremely 'sophisticated disguise for the role of women.'[20] Although a household includes several individuals with differing needs, tastes and preferences, he suggests all neo-classical theory holds it to be the same as an individual. Individual and household choices are, for all practical purposes, interchangeable, and the household is taken as a basic decision-making unit. The obvious problem as to who is participating in the decisions is ignored. Galbraith observes that, in practice, it is usually the man who makes the decisions, and the woman who implements them. 'The household, in the established economies, is essentially a disguise for the exercise of male authority.' Such authority derives from the receipt of income: '. . . in a society which sets store by pecuniary achievement, a natural authority resides with the person who earns the money. This entitles him to be called the *head* of the family'.[21]

The model described is imposed on quite different residential and economic patterns in Third World countries. Census, survey and other population statistics are careful to identify a 'head of household', however arbitrarily that household unit may be defined. Some approximation of the Western pattern of the nuclear family is usually used (parents, children and perhaps grandparents) although it does not adequately describe the great variety of residential and working patterns to be found throughout the Third World which, in many cases, show great flexibility in terms of gaining and losing members, not necessarily close relatives. The idea of a fixed unit called a 'household', a family group residing in a fixed grouping in one house or at most a group of houses, is a figment of the statisticians' imagination. It does not conform particularly well to the situation in Western countries, but is particularly inappropriate and arbitrary in terms of the reality of Third World people. Even more arbitrary is the identification — again according to criteria fixed by the statisticians rather than the people being surveyed — of a household 'head' on whom development is concentrated as if that 'head' represented the 'household' and therefore enabled it to be neatly categorized as if it were an individual. All production attributable to members of the arbitrarily defined 'household', for example, will be seen as pertaining to the individual.

It is in consequence only a short step to defining the 'head of household' as the financial supporter of the household, according

to the Western stereotype of the breadwinner, and as the productive member of it supported by a vague 'family labour'. All
other members of the household are then defined as 'dependants'. On paper, therefore, of the more than 1,500 'dependants'
per 1,000 active people in developing countries, 930 are
females.[22] All that this kind of statistic indicates is that working-
age women, arbitrarily excluded from the concept of participation in the labour force, are equally arbitrarily dumped by
statisticians in a miscellaneous category with children, old
people, and the sick and handicapped. This neatly obscures the
fact that women in fact bear heavy responsibility for supporting
genuinely dependent people.

In many cases where women are officially classified as
'dependants' of a household head, it is clear that in fact they
play a crucial part in the maintenance of individuals in that
grouping, and that, in some cases, the man classified as the
'head' might more accurately be described as a dependant from
the point of view of productive activity. For example, indigenous women farmers of Bolivia, Peru, Ecuador and Mexico frequently provide the major source of family income, combining
production with long-range trading in farm products and handicrafts.[23] A similar pattern can be observed in much of sub-
Saharan Africa, where women support themselves and their
children as well as offering food to a husband. A small sample
from the Bamenda indicates that the women contribute some
44 per cent of the family's gross income. In a sample of Yoruba
farmers, only five per cent of the women receive all their
requirements from their husbands, as compared to 20 per cent
who do not receive anything. The majority are self-employed
on their own land and in trading, and while receiving some
support from their husbands also work on their crops and contribute various services.[24] Most of the women are providing
food, clothing and cash for their dependants.

If there is a man in the 'household', it is virtually automatic
that the planners will define him as the 'head of household'.
Women can only be counted in this category if there is no
likely male candidate, for example in the case of unmarried
women (including widows, divorced, separated and never-
married women), particularly those with children and aged
dependants. In the case of married women living separately
from their husbands — as in polygamous and polyandrous
marriages, for example — classification is inconsistent and
arbitrary. This is also true where a woman's husband is absent
as a migrant worker. The fact that the man identified as 'head

of the household' does not actually reside in that 'household' or contribute to its subsistence may well be ignored in the search for a male head. In a study of a Kenyan project, for example, it was observed without a trace of irony that 'survey data show that 30 per cent of the family heads were absent . . .'.[25]

Even with the strong preference of statisticians for locating a male 'head', one out of three 'households' in the world today are run by women without men present. This is more pronounced the less 'developed' the country and the poorer the social stratum. Female-headed households in the United States, for example, account for just under 20 per cent (mostly among poor black families), while in parts of Latin America it is as high as 50 per cent. The number of officially female-headed households is growing. In Africa this is partly as a result of second and subsequent wives being declared unmarried in some countries, although mainly because of male migration. A similar phenomenon is apparent in Asia, where one factor is the breakdown of the extended family system which involved the re-absorption of bereaved or divorced people into another family network. In Latin America, it is often the women who migrate to the towns, with their children. In many rural areas, patterns of seasonal labour for men result in a series of temporary liaisons, with women and their children forming the only permanent grouping that could be characterized as a 'household'.[26]

The practice of arbitrary definitions of women's and men's roles within a 'household' is far from being on the decline. The current surveys in Africa indicate the extent to which these stereotypes are continuing to determine development planning. One example is an extremely detailed survey of labour productivity in rural areas of Zambia, by the University of Nottingham and the University of Zambia. Two kinds of 'cultivator' are defined, the 'farmer' and the 'villager'; however, these categories seem to be for men only (despite the fact, mentioned in passing, that women do much more of the farming); women can only be defined as villagers' or farmers' wives, as in: 'Farmers tend to have larger families, more wives . . .'.[27] The report, because of its identification of women as 'farmers' wives' and men as 'farmers' (or 'villagers') is full of contradictions and confusions. However, at least it gives enough specific attention to women's farm work to make it clear that women, however mis-defined, are in fact farmers of major importance.

Another report in this series, a work by Bessell on the labour requirements for crops and livestock, manages to obscure almost completely the work done by women and men respectively. For some unmentioned reason he takes the idea of a household head, whom he calls 'farmer (male or female)' and contrasts this person with the dependant 'wife or husband'. The categorization is quite arbitrary and never properly explained; for example, an unmarried woman is not counted as a 'farmer' even though she can hardly be fitted into the dependant category either. Armed with this system of ranking he then finds, after much labour, that people in the dependant category work more hours than the 'farmers'.[28] This presumably reflects the fact that he would have put many more women into the 'dependant' category, and many more men into the 'farmer' category and women work considerably longer in agriculture than men. The whole exercise is an illustration of the obfuscation that results from seeing women in agricultural communities as 'dependants' who can be categorized in degrees of dependency according to their marital status, and the stereotype that this evokes for a Western researcher or planner.

The second case involves the model scheme for agricultural development in Africa for the World Bank — which is likely to duplicate the approach in future projects. This is the Lilongwe Land Development Program (LLDP) in Malawi.[29] The system of classification at LLDP, a variation on the 'head of household' theme, is the basis for its strong bias towards men in the distribution of a variety of development inputs. The program officials deal only with one member of a 'household', who is designated as the 'grower'. All other members of the household are arbitrarily consigned to the limbo of 'grower's wife' or other dependant.

Approximately 70 per cent of the growers were men, and 30 per cent women, in the sample survey of 1970/71. The relatively high proportion of women, in a system where 'when the husband of a married couple is present he is almost invariably selected as the grower', is an indication of the arbitrary character of the definitions. Thus, one of the wives of a polygamous marriage would be excluded from the list of growers while the other (or others) would be counted as growers in their own right, although there is no evidence offered to indicate that the time spent in farming, or the responsibility involved, was substantially different. Similarly arbitrary distinctions are made in the case of women in monogamous marriages (75 per cent of all marriages) according to the presence or absence of their husbands. One is asked to accept that a migrant worker instantly becomes the 'managerial decision maker' merely by arriving in a house where the woman is a working farmer. Given the inflexible register of 'growers' and the constant coming and going of men, it is obvious that a number of 'households' would, at any one time, be classed as having a man 'managerial decision maker' who is not involved in the farm-work at all. Since women are excluded from the category of 'grower' by the project officials' policy of nominating any resident man as the grower, the majority of married women with land rights of their own (which are acquired on marriage in the Chewa traditional system) and farming that land in order to support their children, are made into non-persons as far as LLDP farming policy is concerned. The assumption is that the presence of a man, however erratic, removes all responsibility from the woman in planning for the entire agricultural year and specifically in farming her own land. '. . . female growers are a sub-sample of the total female population, which for one reason or another do not have a husband present to manage their gardens.'

The invisible woman again

The classifications are of much more than academic interest; they provide the framework for re-allocation of women's land to men by project officials, in a land registration scheme discussed in Chapter 6, as well as the basis for identifying recipients for project services. LLDP's Extension and Evaluation Sections have compiled 'Village Grower Registers', which purport to list the names of all the people who possess land in which they are the 'managerial decision makers' — which, as already noted, will invariably be a man if one can be identified. If necessary this will be done by asking a polygamous husband to nominate the 'household' to which he wishes to be officially attached. These Village Grower Registers '. . . provide an extremely valuable census of growers in the area, and, more important, they form the backbone of both Extension and Credit Administration in the Project'.

The categorization of recipients for extension, credit and other services, has been made even more male-exclusive since its inception, with the obvious implications for women's access to these and related services. In the 1970/71 survey, a woman was regarded as a 'grower', i e a 'household head in effect', if she had a polygamous husband who spent most of his time with another wife in her house. In a 1973/74 survey, however, all the wives of a polygamous husband who lived in the same village were classified as one 'household', an example of the arbitrary nature of the 'household' category; the husband was then

designated as 'household head'. Since 44 per cent of the polygamous husbands had both or all their wives living in the same
village, and since polygamous marriages accounted for the
largest category of female 'growers' (over one-third of the total),
this means a substantial reduction in the number of women
officially recognized by the LLDP as farmers. Men would now
comprise three-quarters of all registered 'growers'.

In confining their attention, and directing their questions, to
men only, statisticians and planners produce the same kind of
systematic bias as has been observed in the anthropological
record. Men are likely to provide information, and interpretations of the community's economic activity, which reflect their
own labour inputs and their own area of interest. This is all the
more pronounced because of the well-established tendency of
survey respondents to produce answers that they think the
questioners want or expect to hear.[30]

A check on the reliability of Indian census data illustrates
this point. The exercise revealed that among other biases, the
Census tended to understate the participation of women in
agricultural work, based as it was on information given by the
men of the 'household'.[31] The Gokhale Institute, also in India,
found that attempts to train farmers (presumably the men
farmers) and their literate sons to keep accounts were unsuccessful; the errors and omissions were greatest with regard to the
use of so-called 'family labour', namely women's and children's
work.[32]

It is not only the work of women and girls that may be forgotten by male respondents, but their very existence may be
overlooked. A number of studies of the accuracy of village
population registration have been undertaken, some of which
have found the registers to be very inaccurate, particularly as
regards infant females.[33] Where infant mortality is high children
may well be forgotten, particularly the least valued children, the
girls. Various studies in India, for example, show systematic
under-enumeration of female births and deaths.[34]

It is very striking that, in recent attempts to improve surveys
and micro-level 'village studies', a number of problems have
been identified that clearly relate to the problem of omitting
women and their work from the data base, without the authors
making the connection at all. Michael Lipton and Michael
Moore, in their review of village study methodology, identify
several such problems: the misleading use of monetary measures
of wealth or poverty, with little mention of production for
auto-consumption or barter;[35] the importance of privacy in

interviewing;[36] the failure to arrange interviews at a time suitable to people, on the false assumption that they were unemployed;[37] and the stereotyping of villagers by researchers, which is actually 'an attempt to justify methodologically-unsound procedures and a failure to establish rapport or to introduce checks on responses.[38] All these methodological mistakes are of particular relevance to village women, since they are readily stereotyped by researchers as 'domestic', assumed to be 'unemployed', and are involved mainly in production for auto-consumption or barter, with varying degrees of involvement in informal marketing. Yet Lipton and Moore recommend the timing of surveys according to what they claim to be periods of 'total inactivity' coinciding with the hot season in many developing countries[39] — although in fact this is the case only for men. Having advocated privacy, they abandon this principle in terms of women and men wishing to keep things private from each other (as is frequently the case with financially dependent women who manage to accumulate a small amount of cash), suggesting that 'only' the respondent's nuclear family be 'within earshot' during the interview.[40] Here, they follow the conventional line that 'household economy and structure' represent the smallest unit which it is necessary to examine — even in the most detailed micro-study of a single village.[41] Women are hardly mentioned as respondents, and there is certainly no hint that they be interviewed in the fields, the location suggested for questions about food production. Their only role, it seems, is as 'women of the household' (with the domestic connotation involved in this usage) who are to answer a few questions about food consumption, in interviews in the home.[42] In this context, then, it is hardly surprising that Lipton and Moore advocate an all-male survey team: 'The collection of information on current population structures does not usually require the services of female interviewers'.[43]

The ways in which the data base for planning conceals women's participation, particularly in agriculture, are so diverse that new examples spring up all the time. There is, for instance, the use in agricultural economics of various man-units: 'man-hours', 'man-days', 'man-land ratios' and the rest. Although a superficial observer might assume that 'man', in this academic sense, would probably mean 'person', in fact the convention is quite the opposite, with only males being counted as working one hour when they work one hour. Women may, for example, be allocated two-thirds of an hour when they work one hour, and children one-third. Women, then, are arbitrarily seen as

0.66 of a man-unit. The effect of this is by no means academic. In the LLDP records, for example, the estimated size of the family labour force in a sample survey (i e the proportion of the cropping season that each family member was resident at the holding, and the proportion of working time spent there) was as follows:[44]

	Males, aged 15-64	Females, aged 13-64
Number of family members recorded	1,176	1,124
Estimated family labour force	802	1,057

There were slightly more men than women present in the working-age group, in spite of the anomaly (frequently found in these kinds of surveys) of categorizing females and males by inconsistent age-groups, including a higher proportion of females in the working-age group. However, the actual labour force was made up significantly more by women working (almost one-third more than men). In terms of the arbitrarily defined 'man equivalents', however, the female labour force is reduced to 708, which is less than the male input. Such manipulation of the statistics distorts the entire system of labour-force calculations. For example, in reducing average family size of 4.9 to two 'man units': 'With a conversion coefficient [sic] of females over 13 years equivalent to 0.67 man units, and females under 13 years and males under 15 years equivalent to 0.33 man units, the average full time man units per holding was 2.0.'[45] In effect, it would seem that a woman plus children helping her is reduced to one, or less than one 'man-unit'. The most important result is to minimize the agricultural work done by women by reducing it to a fraction of a 'man-unit'.

Monica Fong, a statistician with FAO, has pointed out that the whole area of studies of working life and the construction of tables of working life to measure a person's expected years of work, is an illustration of the inadequacy of the conventional systems of measurement. For example, the traditional methodology used in the construction of tables of working life for men proved almost unusable for women — even with numerous adjustments. In order to estimate women's working life, a new methodology had to be devised, and this in fact was found to offer several advantages over the old in estimating the working life of men.[46]

Cost-benefit analysis

Just as an argument, even with flawless logic, is only as good as the assumptions on which it is based, so the most rigorous and sophisticated piece of quantitative analysis will be no more reliable than the data base on which it is founded (and the assumptions on which data collection proceeds). There is a growing tendency in the World Bank and elsewhere to use cost-benefit analysis, for example, to make decisions about development projects, programs and policies. Such analysis, given serious distortions and omissions as regards women and specifically their labour input, is likely to increase the problem of women's invisibility by putting a gloss of apparently objective analysis on a biased collection of data. Analysis involving 'man-units', as described above, will completely obscure the importance of women to the labour input. Various strange effects may be produced; for example, in the quest for a uniform ranking system for projects,[47] the reduction of women to a fraction of a man-unit could mean that a project involving more men than women would be given higher ranking than an equivalent or better project involving more women. The option of showing separately the costs and benefits of a given set of projects to different groups in society, and letting the policy-makers decide on the relative weights to be given to each, is only a theoretical possibility rather than actual practice.[48] And, as Pearce observes, '. . . practical studies are frequently not able to utilize the refinements suggested by theory, most often because of lack of data'.[49]

It is one of the general problems of cost-benefit analysis that it is concerned only with total costs and total benefits without regard to the distribution of these; perhaps its weakest assumption is that those receiving benefits from a project will somehow compensate those who suffer costs. Mishan argues that such compensation is not in fact paid:

> 'A project admitted on a cost-benefit analysis is . . . quite consistent with an economic arrangement which makes the rich richer and the poor poorer. It is consistent also with transparent inequity: irrespective of the income groups involved, the opportunities for increased profit or pleasure involved by the new project may inflict direct and substantial injury on others.'[50]

Frances Stewart goes further, arguing that cost-benefit analysis is not value-free, but systematically favours certain sections of the community which wield political power — and these will normally be the higher-income groups.[51] Certainly if time saved is evaluated in terms of the market price of an individual's work, then time saved by relatively rich people will appear to be of greater benefit to the community. Also people earning cash or assumed by the analysts to have an 'opportunity cost' attached to their labour-time (usually the men), will be considered of disproportionate importance, with benefits to them appearing of greater significance than the same benefits to others. 'Time saved' often accounts for between 80 and 90 per cent of the benefits accruing to transport projects, for example, and this is valued at a proportion of the income of the individuals concerned.[52] The benefits of some health programs are measured in terms of the increases in GNP which result from earlier returns to employment.[53] Thus a program directed at diseases afflicting mainly men would appear more desirable than one directed at maternal and child health, for example. The same applies to the value of an individual life. Mishan observes in passing that men's lives are deemed much more valuable in cost-benefit analysis than women's.[54]

The problem applies particularly to social cost-benefit exercises. Stewart has pointed out that the measurement of benefits (or social welfare) generated by a project cannot be separated from the distributional consequences: '. . . the measure depends on the point of view adopted'.[55] Different class and interest groups have their own common preferences and values. To each set of values, there corresponds a set of shadow prices — i e, those prices which would contribute most to the objectives. 'If used for project evaluation a different set of projects would be chosen according to whose values, and hence which shadow prices, were being used.'[56] She goes almost as far as describing social cost-benefit analysis as a confidence trick. Government values (and priorities) are accepted uncritically as some kind of synthesis of the national interest, although in reality a government represents primarily the interests on which it depends for its power.[57]

> 'Social Cost-Benefit analysis, in so far as it implies that social welfare maximization or national welfare maximization is meaningful (and also possible) in conflict societies, is highly misleading and sometimes dangerously so, since it dresses up one set of activities — those of taking the objectives of one section of society, normally those represented by the Government, and showing how they may be more efficiently fulfilled — as another, that of maximizing the benefits to society.'[58]

It is fairly widely recognized that certain kinds of divisions within society provide possible sources of conflict affecting the value of cost-benefit analyses; Stewart, for example, mentions urban/rural divisions, generational conflict, race, religion, tribe and caste.[59] Layard describes concern about distribution among American economists as being directed only at income groups, age groups, racial groups and geographical areas.[60] Weisbrod's list, unusually, includes sex as one of the factors in distribution, although he does not elaborate on this.[61]

Virtually no reference is made in the literature to the ways in which women's work is valued (or devalued) by cost-benefit analysis. Gittinger, in a rare exception, states that women are assumed to be completely unproductive if there is too little land available to a farming family. Although not stated explicitly, it is clear that women's non-farm work is automatically classified as unproductive, just as it is excluded from GNP calculations. In fact women are seen as a complete liability in such situations:

> 'A common example where a "wage" is paid even though no productive work is available is found in the case of family labour. Older children and the farmer's wife [sic] will most certainly be entitled to a share of the family income, even if the family farm is too small to give them an opportunity to be productive.'[62]

Thus in the case of large numbers of Asian women, for example, who work extremely hard to produce essential goods and services without which the family cannot survive, the marginal value product of that labour would be deemed zero. Magnanimously he adds: 'More recently, professional opinion has swung to the view that the marginal value product is not quite zero, but often very close to it'.[63]

Conclusion

As Galbraith has observed, the concealment or disguise of women's work serves a very important function; since 'what is not counted is usually not noticed',[64] planners are able to assume that, in a literal sense, women do not count.

The identification of the entire population as 'men', in the most ambiguous usage of that word, is almost total. It is a short step from calculating people's work in terms of 'man-units' to discussing what to do for the men. Even the most perceptive observers have the habit. Raymond Apthorpe and Fiona Wilson assert, 'Many writers agree that for development to take place a particular kind of man is necessary . . .'.[65] Robert Chambers suggests that '. . . the opportunities and problems of managing communal natural resources involve a dimension of human management — that is, of the management of men in organizations and in communities'.[66] Attempts by large development organizations on the international plane to break through the jargon and the GNP to reach the small farmer are equally addicted to the man. Robert McNamara, in his October 1976 address to the Board of Governors of the World Bank, announced what was to be a radically new policy of aiding the disadvantaged: 'We meet this year against the background of growing recognition that equality of opportunity among men, both within nations and between nations, is becoming a critical issue of our time.' Elizabeth Reid comments that he has failed to perceive the most glaring inequality of all in his agency's approach to people in the Third World. His use of language, she observes, '. . . mirrors a perception of reality of a world where many men are, of course, marginal but where the women, if perceived at all, are perceived as marginal to the marginal'.[67]

Excluded, then, from the mainstream of economic life with which development planning is primarily engaged, concern about women, which since International Women's Year has been gaining ground in some development agencies, is strictly marginal to the day-to-day preoccupations of planners. The solution, as with many other disadvantaged groups, is to categorize women as 'social problems', requiring the establishment of special welfare-orientated projects that can sometimes be attached at the margin of an economic development project, and sometimes be set up as separate 'women's projects'. It is these projects and programs which are examined in the next chapter.

References

1. Harold Brookfield, *Interdependent development*, London, Methuen, 1976, p 26.
2. *Ibid*, pp 23, 26.
3. *Ibid*, pp 26, 40
4. Interviews by the author with officials of the United Nations, New York.
5. United Nations Statistical Office *Report of the expert group on welfare-oriented supplements to the national accounts and balances and other measures of the levels of living*, UN document ESA/STAT/AC.4/5, 15 April 1976, pp 2-3.
6. *Ibid*, p 3.
7. See e g US Department of Health Education and Welfare, *Economic Value of a housewife*, Research and Statistics Note No. 9 — 1975, DHEW Pub No (SSA) 75-11701, August 28, 1975. Also John Kenneth Galbraith, 'The economics of the American housewife,' *Atlantic*

Monthly, Vol 232, No 2, August 1973, pp 78-83.

8. Galbraith, *op cit*, p 79.

9. *Ibid*, p 83.

10. Carolyn Shaw Bell 'GNP: meaning behind figures,' *Christian Science Monitor*, 15 April 1976.
    The major practical objection to including the value of women's production in GNP is in fact the difficulty of determining a reasonable market price for that production. See e g United Nations Statistical Commission, *The feasibility of welfare-oriented measures to complement the national accounts and balances: report of the Secretary-General*, UN document E/CN.3/477, 17 February 1976, paras. 29-71. Also US Department of Health, Education and Welfare, *op cit*, and UN Statistical Office, *Report of the expert group*, paras. 11-15.

11. Hans Singer and Richard Jolly, 'Unemployment in an African setting,' *International Labour Review*, Vol 107, No 2, February 1973. Reprinted in *The Pilot Employment Missions; and Lessons of the Kenya Mission*, IDS Communication 111, n d.

12. Gustavo Pérez-Ramirez, 'Unveiling women in statistics,' *Populi*, Vol 5, No 1, 1978, p 17.

13. *Ibid*, p 21.

14. *Ibid*.

15. Singer and Jolly, *op cit*.

16. *Ibid*, p 3.

17. Cherry-Lynn S. Ricafrente, 'Accelerated manpower development program in the Philippines,' in National Manpower and Youth Council, *Background papers on manpower development*, Manila, National Media Production Center, 1973, p 5.

18. See e g William Allan, *The African husbandman*, Westport, Conn, Greenwood Press, 1977. Also Martin Upton, *Farm management in Africa: the principles of producing and planning*, London, Oxford University Press, 1973.

19. F Jurion and J Henry, *Can primitive farming be modernised?* (translated from the French by Agra Europe), London, Institut National pour l'Etude Agronomique du Congo, 1969, pp 48-49.

20. Galbraith, *op cit*, p 79.

21. *Ibid*, pp 79-80.

22. Pérez-Ramirez, *op cit*, p 17.

23. Teresa Orrego de Figuerea, 'A critical analysis of Latin American programs to integrate women in development,' in Irene Tinker, Michele Bo Bramsen and Mayra Buvinic (eds), *Women and world development*, New York, Praeger, 1976, p 47.

24. Ester Boserup, *Woman's role in economic development*, New York, St. Martin's Press, 1970, pp 42-43.

25. Uma J Lele, *The design of rural development: lessons from Africa*, (World Bank Research Publication), Baltimore, The Johns Hopkins University Press, 1975, p 55 (published for the World Bank).

26. See Irene Tinker, 'The adverse impact of development on women,' in Irene Tinker *et al* (eds), *Women and World Development, op cit*, pp 22-28. See also United Nations documents for the Mexico Conference of the International Women's Year, 1975.

27. Universities of Nottingham and Zambia Labour Productivity Investigation (UNZALPI), *Some determinants of agricultural labour produc-*

tivity in Zambia, Report No 3, Lusaka, University of Zambia; Sutton Bonnington, Leicestershire, University of Nottingham, 1970, p 12.

28. J E Bessell, *Labour requirements for crops and livestock*, Report No 4 of UNZALPI.

29. The sources of information on this project are confidential.

30. See e g B A Phieps, 'Evaluating development schemes: problems and implications, a Malawi case study,' mimeo, Dar es Salaam, East African Universities Social Science Council, Annual Conference, December 1970.

    A Tiv informant comments on this question: 'When I read what the white man has written of our customs, I laugh, for it is the custom of our people to lie as a matter of course to outsiders, especially the white man. We ask, "Why does he want to know such personal things about us?" ' — Quoted in Frank A Salamone, 'The methodological significance of the lying informant,' *Anthropological Quarterly* Vol 50, No 3, July 1977, p 117.

31. Michael Lipton and Mick Moore, *The methodology of village studies in less developed countries*, Brighton, Institute of Development Studies, Discussion Paper No 10, 1972, p 51.

32. *Ibid*, p 54.

33. *Ibid*, p 52.

34. *Ibid*, p 70.

35. *Ibid*, p 66.

36. *Ibid*, pp 67, 100, 102.

37. *Ibid*, p 99.

38. *Ibid*, p 67.

39. *Ibid*, p 99.

40. *Ibid*, p 100.

41. *Ibid*, p 102.

42. *Ibid*.

43. *Ibid*, p 69.

44. Adapted from LLDP data.

45. *Ibid*.

46. Monica Fong, 'Women and development: the need for a really innovative approach,' *Ideas and Action Bulletin*, No 112, 1976/5, p 22.

47. See e g Albert O Hirschman, *Development projects observed*, Washington DC, Brookings Institution, 1967, p 174 and ff.

48. See Richard Layard (ed), *Cost-benefit analysis*, Harmondsworth, Penguin Books, 1972, p 60.

49. D W Pearce, *Cost-benefit analysis*, London, Macmillan, 1973, p 68.

50. E J Mishan, 'What is wrong with Roskill?', in Layard (ed), *op cit*, p 453. Chapter 16 also has a discussion of the importance of distributional effects.

51. Frances Stewart, 'A note on social cost-benefit analysis and class conflict in LDCs,' *World Development*, 1975, Vol 3, No 1, pp 31-39.

52. Pearce, *op cit*, p 74.

53. *Ibid*, p 57.

54. E J Mishan, 'The value of life', in Layard (ed), *op cit*.

55. Stewart, *op cit*, p 36.

56. *Ibid*, p 33.

57. *Ibid*, p 35.

58. *Ibid*, pp 36-37.

59. *Ibid*, p 34.
60. B A Weisbrod, 'Deriving an implicit set of governmental weights for income classes,' in Layard (ed), *op cit*, p 403.
61. *Ibid*, p 427.
62. J Price Gittinger, *Economic analysis of agricultural projects* (The Economic Institute, International Bank for Reconstruction and Development) Baltimore, The Johns Hopkins University Press, 1972, p 41.
63. *Ibid*, pp 42 and ff.
64. Galbraith, *op cit*, p 79.
65. Raymond Apthorpe and Fiona Wilson, *Social Planning*, Brighton, Institute of Development Studies, Communications Series No 58, 1970, p 21.
66. Robert Chambers, *Two frontiers in rural management: agricultural extension and managing the exploitation of communal natural resources*, Brighton, Institute of Development Studies, 1975, p 1.
67. Elizabeth Reid, 'The forgotten fifty per cent,' *Populi*, Vol 4, No 1, 1977, p 24.

Chapter Five
The new segregation in development projects

In the name of integration

Having effectively excluded women from the economic base, by oversight or by reducing them to fractional 'man-units', planners can quite logically argue that women, who have certain moral or 'social' claims, be given special consideration over and above the economic projects of the mainstream. Thus, the question may be put how planners can incorporate a 'women's element' into cost-benefit analyses.[1] At the operational level, women are often allocated a special section of a development institution, and a special set of programs tailored to the 'social' role assigned to them by the planners. This tendency becomes more pronounced as demands for the 'integration of women into development' become more urgent, as has generally been the case since International Women's Year in 1975. Many Third World countries and international development agencies have set up special offices to take care of women's concerns, in line with such arguments as those put forward in a resolution of the International Labour Conference in 1964, which requested all member states to consider establishing a central unit for co-ordinating activities for women workers. Since then many countries have set up women's bureaux, or women's departments within Ministries of labour, social welfare or social security.[2]

Where women reach the highest rank — as in Britain, Argentina, India and Sri Lanka — they are backed by virtually all-male governments and institutions and, as heads of state, relate to women no differently than their male predecessors. At the same time, most of the women who make their way to the higher administrative and political levels are often channelled in the direction of these specialized, generally 'social' activities. As Kathleen Newland observes: 'Where they do hold cabinet or sub-cabinet positions, women tend to be concentrated in "soft-issue" areas — health, welfare, social services, and so forth.' She quotes Françoise Giroud, the former Secretary of State for Women's Affairs in France, as commenting:

'It seems to me interesting to note that when, for the first time, four women are members of the French Government, what responsibility is committed to them? Hospitals, children, prisoners and women. Nothing, in short, that might frighten men and bring them to think that women may invade their territory.'[3]

The case of Bangladesh is an interesting example of how 'women's issues' are kept marginal. Considerable efforts have been made at various levels — governmental, non-governmental and among some United Nations agencies — to integrate women into development planning. Reports by Adrienne Germain of the Ford Foundation suggest that a relatively favourable situation exists in Bangladesh from this point of view. Structures created to promote women's interests include a Special Assistant to the President, a Women's Affairs Division and the JMS (Jatiyo Mohila Sangstha), a national women's organization set up by the Government. 'The high level political commitment and interest is, in fact, unusually favourable', and achieved within a relatively short time. As with most government intervention, however, the structure created for women is:

'. . . controlled by male civil servants and élite women with a traditional social welfare orientation. It is . . . remote both from the majority of women it is meant to serve and from the major operative ministries (education, labor and social welfare, population control and family planning, establishment, rural development, agriculture) whose policies it should influence.'[4]

Position papers, although improved since 1977 when Germain first reported, are confined mainly to generalities, with few hard facts especially on rural women. Laundry lists of recommendations omit key items: for example, there are no plans to evaluate, let alone attempt to reduce the extent of male bias in plans now being made for the 1981 census; and there has been no systematic review of government policies in agriculture, education and employment to identify sources of discrimination.[5] A panel of Bangladeshi economists describes the 1978-80 National Plan as follows:

'The Plan document by dubbing women's programs as "welfare programs" and relegating them to one of the two subsectors of the social welfare sector, has faithfully mirrored the male dominated society's attitude towards women . . . they have been treated at par with the marginal groups . . .'[6]

The relevant section allocates 55 per cent of the proposed budget for social welfare to the section in the ministry which deals with women; they explicitly describe women as being 'in a state of total dependency' for which the best available solution is training and production centres for handicrafts.'[7]

The situation has improved somewhat in the last two years. For example, the first batch of female agricultural agents has been trained, and there is a research unit in the Ministry of Agriculture that gives particular attention to women. These are important first steps. Nonetheless it is still true that most of the planners lack any understanding of the realities of most women's lives in Bangladesh. They pay little attention to the bulk of women's work which is agriculture-related (processing, storage and preservation of the major food crops). As Germain puts it, 'women have been discovered'; yet, although international agencies seem anxious to spend 'rather large sums of money' on women's programs, they tend to restrict their 'women's funds' to the Government's social welfare programs. These contributions are to be 'segregated from their major development programs in agriculture, employment, development planning, education and so on.'[8]

International organizations: a woman's place is in home economics

Of the United Nations bodies, two key organizations will be examined with regard to their treatment of women in the organization of their field projects and programs, and the overall organization of their work. They are the United Nations Development Program (UNDP), which — in theory at least — is the controlling organization, banker and field representative of the entire United Nations 'family'; and the UN Food and Agriculture Organization (FAO), the largest of the UN agencies and the one specializing in rural development and agriculture.

A report on UNDP-assisted projects involving women was prepared in 1974 by Thanhdam Truong. A careful search of all current projects and programs, although not definitive because of the failure of Resident Representatives to report the breakdown of participation by sex, identified a number which seemed to be specifically directed at women; a further group of project reports mentioned women's participation. Most of the projects related to 'home economics' and its sub-branches, such as nutrition and 'child-rearing'. A few others involved training women, mostly in a narrow range of jobs stereotyped by Western culture as being for women (e g hotel work in Singapore, 'clerical, secretarial and sale [sic] personnel' in Senegal, work in textile factories in Thailand (their only access to the Small Industries Service Institute), and nurse/midwives in Democratic Yemen. A third category consisted of health and welfare projects for women and children. Overall, the number and size of projects for women, as reflected in Thanhdam Truong's report, was minimal in relation to the overall level of UNDP involvement in the countries concerned.[9] This is borne out by an example from Africa: the UNDP Resident Representatives report on development assistance to Malawi for 1976, although it does not identify the degree of participation by women and men respectively, shows only one program, now terminated, among all externally-funded operations, which is likely to have been for women. This is the UNDP/FAO 'home economics' project (MLW/68/004) for which a total of $4,000 was allocated in 1976. This compares with total allocations for the year of $12.5 million.[10]

As Truong concludes in her report on UNDP, '. . . it seems that development policies are being more influenced by a segregation of sex roles than an understanding of the specific needs of women'. Technical assistance to women in agriculture was 'marginal in some of these countries and non-existent in others'. Almost exclusive concentration was being given to training women in home economics.[11]

The problem of imposing Western stereotypes of sex roles on Third World women is by no means disappearing as UNDP becomes more involved in debates about women's role in development. On the contrary, the stereotype forms the basis for UNDP's activity on this issue. In charge of all efforts within the organization to promote women's involvement in field projects is Ulla Olin, Principal Officer in the Program Policy Division of UNDP and responsible, among other things, for drafting the guidelines for field officers for use in their work with women. She has published a paper on the subject which leaves no doubt as to her faith in the stereotype of women's place being in the home. Basing her arguments on the situation of women in upper-class suburbs of 'the highly-developed countries' as the pattern for women, which is the only apparently feasible one currently available, she argues that women have had to spend a very substantial part of their adult lives 'in or near the home'. This is seen as an evolutionary imperative based on men's reproductive behaviour which is closely related to 'aggression', whereas women's allegedly involves 'caution and submission'. She concludes that 'family or reproductive behaviour may be assumed to remain the focus of human endeavours, however remote from actual biological reproduction these may appear to be'. Women 'everywhere' play an important role as managers of 'the universal microcosm of family life'. Their 'behavioural inclinations' are quite different from those of men, making them unfit for 'object-oriented, large-scale social organization'. Because of this gulf between women and men, women should be allowed the opportunity to contribute to public life *as women* (Olin's emphasis), rather than as people; any demand for equality with men 'is erroneous and doomed to fail' because of the 'dominant position of men'. It cannot be beneficial for women to insist on having the same rights and duties as men, at least in the world as we now know it. The issue of equal rights, indeed, may aggravate existing social problems by 'intensifying the social competition for status and the like'.[12] While clearly well-intentioned, and making the useful point that women are likely to bring a different approach to management than men which may be beneficial, the whole philosophy behind the argument is exactly the same as that which lies behind the exclusion of women from economic data and the planning based on that data. It illustrates well the implications for development policies of the Western male ideology outlined in Part One: the logical conclusion of this argument is the segregation of women into special projects, which has already been observed to be the basis of UNDP's approach.

Progress towards the stated goal of 'the integration of women in development' looks splendid on paper; but the practice is very different. Much emphasis is placed on UNDP's support for a women's program in Upper Volta, and for the African Training and Research Centre for Women, which is attached to the Economic Commission for Africa (ECA). However, the project in Upper Volta, used as an example of UNDP's dynamism on this issue,[13] has failed to affect the mainstream of work in UNDP field operations in that country, which remain male-centred.[14] The African Training and Research Centre for Women, being in the 'Social Development' category,[15] has a very low priority. Although linked with the ECA it receives little financial support from it on a regular basis, and depends almost entirely on a variety of special contributors.

There is no requirement in UNDP that project designers consider women as integral elements; the Policy and Procedures Manual, the field officers' bible, merely requests that 'when applicable', they describe under the low-priority heading 'Special Considerations' how the integration of women in development is to be pursued.[16] The extremely marginal impact of this kind of 'special considerations' approach is obvious: planners have inumerable serious issues to consider in writing up a project document, and the relegation of what is claimed to be the 'integration' of women into a category that is not usually used, right at the end of the document, actually represents the segregation of women out of the main projects. A simple example may illustrate the point. Botswana sought UNDP assistance for a vocational training program to train instructors at the Botswana Training Centre, and a centre is also to be built for training in electrical work, building, fitting, welding and plumbing jobs. A report prepared for the United Nations Office of Technical Co-operation delicately points out that 'no special facilities have been designed for girls in the architectural plan of the centre';[17] in other words, there are toilets for boys but not for girls. The same argument as that used against training women as agricultural extension agents is thus being set up: 'no facilities'.

As reported by UNDP's representative at a regional conference on women in development held in Nouakchott, Mauritania, in 1977: 'Rather strong assertions were made at the Nouakchott Conference regarding UNDP inertia in the area of women in development'.[18] The organization is good at the rhetoric of integration, its Administrator Bradford Morse having made a number of positive statements about 'integrating' women in development; the reality, however, is that they are virtually ignored.

The FAO, in contrast, has no inhibitions about where women's place is to be. This organization has a department devoted to home economics which, apart from providing a ghetto where many of the FAO's professional women congregate, has official responsibility for all questions relating to women in the entire range of FAO's activities in agriculture and rural development. The head of home economics projects is also consulted on decisions about women in every other context, regardless of her actual expertise in the subject concerned.

In the field, FAO's activities for women are almost exclusively confined to small projects, or often components tagged onto large rural development schemes. They are home-economics based, with emphasis on motherhood, nutrition, sewing and knitting, housework and sometimes handicrafts and voluntary work. The title of these schemes often includes the word 'family', apparently as a euphemism for women, as in 'Programs for Better Family Living' which in practice mean training women in home economics. The Home Economics service of the FAO prepared a major publication for 1975, International Women's Year, on 'The Family in Integrated Rural Development'. While hardly mentioning the word 'women', it can be seen as part of the claim of home economists in the FAO to assert their leadership over this half of humanity: 'A forward-looking, modern, adaptable, home economics program has thus a considerable potential contribution to make to the achievement of the goals of rural development'.[19] (As far as is known, home economists make no effort to involve men in their 'family' programs.)

An inventory on FAO activities related to women, completed in December 1976, shows that 83 per cent of FAO's activities with an identifiable component for women are planned by two divisions: the Food Policy and Nutrition Division (reflecting the cookery and nutrition projects) and the Human Resources, Institutions and Agrarian Reform Division (since this includes the Home Economics and Social Program Service). Most other FAO units reported very few activities which showed evidence of benefiting or involving rural women. For example, the Agricultural Services Division reported activities amounting to only 5 per cent of their total; General Affairs and Information Department reported 2.6 per cent; Forestry Department reported 1.7 per cent; Statistics Division and Policy Analysis Division each reported one activity only; and the Fisheries Department reported none. A total of 212 activities were reported which had some 'potential contribution' for rural women if they could be reorientated in that direction, although there is no sign of this reorientation actually taking place.[20]

What is home economics?

The development of 'domestic science', the American version of 'home economics' or similar variations on the theme, has been described in relation to the domestication of Western women in the early twentieth century in Chapter 1. It is essentially this type of missionary activity which characterizes the enthusiastic proponents of the art in the Third World, using the pretensions of science (or economics, apparently an interchangeable concept here) to impose new standards of housework on the poor benighted savages (immigrants, the urban poor, or ignorant peasants). Some of the younger members of the home economics ghetto, in the FAO and elsewhere, are even more ambitious in their claims to speak for all women than are their seniors. The grounds on which they make these claims are that the 'new' home economics is moving away from the traditional concerns of sewing, knitting, cookery, childcare and the rest, to incor-

porate new elements such as social survey techniques, agricultural innovation, appropriate technology, functional literacy
skills and training related to the generation of income for
women. However, the actual content of many home economics
projects and programs remains solidly housework-based, a result
of the success of the Western stereotype of women as domestic,
which has been incorporated into the expanding education system in Third World countries since the Second World War. The
ambitious new home economists, in fact, have more in common
with their predecessors than they will admit: the domestic
science movement showed the same missionary fervour in its
early days, the same pretensions to know best for all women,
and the same eclectic tastes in scientific fashion.[21]

The post-war period, as already noted in Chapter 1, was the
heyday of 'maternal deprivation' theories, which accompanied
the rapid exclusion of women from important sectors of the
employment market in Western countries and the ideology of
domestic fulfilment (through consumption) that accompanied
it. The post-war establishment and subsequent evolution of
development institutions, as we now know them, were deeply
influenced by the idea that economics were a matter for men,
and that women's concerns were restricted to the domestic
sphere.

It is not the intention here to dismiss altogether the value in
a given situation of the special projects for women; it is, however, much more important to recognize the problems of pursuing this kind of approach as if it were conducive to the 'integration' of women in development when too often it achieves
the opposite.

An argument can certainly be made for special efforts to help
women in the Third World: it would have to involve an un-
biased attempt to understand their problems and needs, and the
evolution of an approach designed to meet these needs. The
main characteristic of the 'home economics' approach, however, is the reverse: it applies a Western concept of domesticity
to Third World women without attempting to understand their
work in any other than a domestic context. There is a strong
prejudice about what women's work is, and if the reality does
not fit that preconceived pattern then a strong element of moral
disapproval is introduced. Women, it is assumed, should spend
more time with their children; should do more washing and
cleaning; should learn new recipes and so on. The ideology of
Western middle-class 'right living' is imposed, in the same way
as among early 20th century immigrants in the United States.[22]

As Adrienne Germain points out, the focus of special projects is all too often peripheral to women's real lives. 'Special projects seem to be weakest in teaching women how to do their routine work more efficiently and productively.'[23] She attributes this to the dearth of information available on what women do, and how; ways in which their behaviour could usefully be modified; and the technologies and other inputs that would increase productivity without displacing labour.[24]

The new generation of home economists claim to be incorporating new elements such as small-stock husbandry, horticulture and certain income-generating activities like handicraft production, which involve a more useful approach to women's work. However, one characteristic of these activities is that they are additional to women's routine work; they usually involve learning new skills rather than applying resources to routine occupations. Given the enormous load of overwork which so many rural women bear — discussed in Chapter 7 — the prospects for introducing new activities are severely limited. Adrienne Germain makes the point that in Bangladesh, women's routine work including food preservation and storage, livestock husbandry and agricultural production can be income-generating if efficient enough to produce a surplus; she criticizes the bias toward developing 'new' or extra activities for women, particularly in craft projects.[25] Bangladesh is the focus of many of the new varieties of 'women's project', and the criticism is generally applicable to the new wave of experimental projects now becoming evident in many other countries.

The domestic factor

It is relevant here to examine the impact in Third World countries of the development agencies' focus on special projects for women. We need to establish, as far as the evidence will allow, to what extent rural women are channelled into home economics, when they do have access to development resources; the scientific basis for the activities; and the actual impact of the projects. Finally, we shall consider evidence relating to rural women's interest in different types of project; and the effect of establishing women's projects on the prospects for integrating women into development planning as a whole.

In assessing the degree to which home economics dominates the education and training available to women and girls, it is important to emphasize that the influence of this subject goes far beyond the individual projects or project components which are offered by such agencies as FAO. These are, as we have seen, heavily concentrated on home economics as an integral principle in the functioning of the agency. Far more influential than these projects for women in developing countries is the introduction of a compulsory 'home-science' element into the entire range of formal and non-formal education, and of vocational training.

Ester Boserup, surveying the issue in the late 1960s, observed that the emphasis in education and training of girls was on 'household' work, mainly cooking, child care, sewing and embroidery. Such subjects took up much of the time in many primary schools; the courses offered under programs of community development and rural extension were largely devoted to them; and even at university level, 'much of the teacher training of women is in fact a training for the role as instructor in home economics and similar subjects'.[26] She also noted that home economics was taught to rural girls as an alternative to the agricultural courses which increasingly were being introduced for the boys in the effort to make education, especially at primary level, more 'relevant'. The same phenomenon of directing men to agriculture and women to domestic subjects, with no-one able to learn the whole range, was observed to be usual in adult training programs.[27]

The Economic Commission for Africa estimates that domestic science constitutes more than 50 per cent of all the non-formal education offered to women. Agriculture and co-operative education are very rarely offered, despite the predominance of women in agricultural work in Africa. Women comprise 100 per cent of participants in home economics as compared with 15 per cent for agriculture and zero for trade and commerce – another area where many women actually participate fully. Men, on the other hand, comprise only 10 per cent of participants in nutrition courses, and zero in home economics.[28] The same phenomenon has been noted on a world-wide scale by Gloria Scott, who points out that in recent years literacy education has become more 'functional', rather than simply providing a general reading skill. Since women's 'function' is defined arbitrarily as domestic, it is no surprise that she observes them being increasingly confined to domestic science, with instruction in foods, health, family planning and other 'domestic' issues. The men are not offered any of these courses.[29]

On the most informal level of education, that of access to various development extension agents, the pre-allocation of women and men to home economics and agriculture respectively is even more pronounced. Uma Lele comments that not only do planners perceive female farmers as if they were Western-style housewives, but they deliberately set this as the development objective:

> 'Thus, the goal of extension services has frequently been not the increase in farm-level productivity of women but rather finding ways to reduce their participation in agriculture through promotion of more homebound activities . . . Too often women's extension programs have been exclusively oriented toward domestic science and home economics.'[30]

She adds that the opposite is achieved to what is intended, since extension programs aimed exclusively at men and their activities have the effect of increasing women's agricultural workload — a major problem to which we shall return in Chapter 7.

The situation can perhaps be illustrated by the case of a particular project relying heavily on extension agents: the World Bank's model project for 'integrated rural development' in Malawi, the Lilongwe Land Development Program (LLDP). Training programs are segregated, with the Development Officer for 'women's training' having 'overall responsibility for all female staff of LLDP' and 'responsible for the content of all women's courses', according to project documentation.[31] The women's training program includes home economics, nutrition, 'needlecraft', childcare, and a range of other domestic skills.[32] Men, meanwhile, receive technical assistance in agriculture and animal husbandry. Female extension agents are given the title of 'farm home instructress' [sic], implying that they are directed at women's tasks performed in or around the home, rather than the field work which occupies much of their time. In many if not most of the families, men are periodically or permanently absent, and women are responsible for the operation of the whole farm, including field crops and livestock; but they are denied access to the extension and training services offered to men.

Even in the rare instances where particular efforts have been made to set up training for women in areas already available to men, it seems that the planners cannot resist inserting a domestic element even if it means eliminating important parts of the main course. In Bangladesh, for example, where women were recruited to begin training as agricultural extension agents, the courses have been changed to delete those on machinery (an innovation with major consequences for women's work) and substitute 'home science'.[33] When women participated in farmers' training courses in Kenya, there was always some instruction in home economics for them, the argument being that this is what the men want women to do.[34]

Far too many of the much-publicized 'women's projects', however, are little more than the old home economics projects, perhaps with a token amount of poultry-keeping or appropriate technology attached to make it look more relevant to rural women. One of the agencies making most noise about women's projects is the US Agency for International Development (AID), which has a legislative mandate, in terms of the 'Percy Amendment' to the Foreign Aid Act of 1976, to integrate women into development. A typical example of their new activities under this provision is a project in Upper Volta, set up after much discussion in 1977. Although this is supposed to be a new departure, it looks very much like the old style of home economics; according to the Project Director, training is to be provided for Agents d'Economie Familiale (domestic economy agents), to upgrade their work in. health and hygiene; sanitation; literacy; 'traditional home economics' and unspecified agricultural or economic activities. The agents are all employed by the regional development organizations, or ORDs, which see women's programs as marginal to their major effort directed at men; out of some 1,500 extension agents, a total of 95 are women.[35]

This reflects the problem inherited by any agency which attempts a new style of 'women's project': the available agents, or potential agents, are firmly set in a home-economics mould. Joy Greenidge, another official of AID who has spent many years in Upper Volta, commented that most of the girls available for recruitment to such training programs have their basic CAP qualification in home economics; training is therefore made 'complementary' to that, with courses in nutrition, health and the rest. A related problem is that the range of training available to rural women has been so extremely narrow that if they are asked what kind of teaching they want, they will often request sewing lessons since this is the only activity that is familiar to them as associated with extension in general. As soon as the prospect of alternatives is realized, the choice is a very different one.[36]

The assumptions underlying the restriction of so much education and training for women to home economics are simple: that women's place is in the home; if it is not, it should be; and men have no productive place there at all. No concessions are made to societies where men and/or boys play an important role in child-care, where they wash their own clothes, prepare food, sew, make household goods, or perform other activities within the range of the home economists. It would be unusual, for example, to find them uninterested and uninvolved in caring for children, whether as their own or their sisters' offspring or, for boys, their sisters and other close relatives.

The myth of maternal deprivation — but what about the fathers?

The concept of 'maternal deprivation', the result of women not giving virtually exclusive attention to their own children in the first few years of life, has already been mentioned in Chapter 1 as a powerful element in the Western male ideology, legitimizing the domestication of women after the Second World War. This was just the time when the institutions and basic ideas of Third World development were being established, and the ideas of Bowlby and his followers have been extremely influential in determining the view that women should be confined to a domestic role, as full-time 'mothers', and that this would automatically exclude them from mainstream development.

Child-care is perhaps the most emotive issue for home economists and the one where local realities are most likely to be ignored. Agencies concerned mainly with children, such as UNICEF and various non-governmental bodies such as the Save the Children Fund, focus almost exclusively on children and their mothers. They often have relatively large amounts of money at their disposal, and are among the most active advocates of special projects for women — mostly those that will reinforce their 'mothering' role. The objection that fathers are also responsible for children is, in this context, a novel idea, just being introduced into UNICEF meetings by Scandinavian delegates, but by no means yet applied to the agency's programs.

The care of children is of course a major concern of women everywhere; however, the need is not for lessons in 'mothercraft' from self-appointed outside experts, but for the means of feeding and maintaining the children. This means generating food and income by their own work in agriculture or other activities. A major finding of a study of poor women in Tamil Nadu, India, was that most of them defined their role 'as a woman' not in the expected Western terms of being a wife and mother, but 'to earn and support the family'.[37] In many parts of the world, the domestic role of women in a nuclear family as promoted by home economists is applicable only to the richer minority. Lucy Mair mentions that in Guyana, marriage is for 'middle-class' people, and seen as pretentious for insecure and poor people.[38] This applies in varying degrees to much of Latin America and the Caribbean, and increasingly to Africa and parts of Asia, where the family is being broken up and women left to care for children alone.

The 'family' model put forward by Bowlby and his followers as the basis for society, with the women in that family confined to child-care and domestic work, is one demanding relatively high and secure income-levels and, in addition, an aspiration by the people involved toward the Western way of life. It is interesting that so many home economics groups, although supposedly aimed at the poorest women, tend to involve the wives of salaried men. Adrienne Germain comments that in women's groups (as in men's groups) organized by development agencies in Bangladesh, participants are often from the better-off families especially at the inception of the group.[39] In Botswana, where a rare attempt at evaluating home economics teaching has been made by Marit Kromberg and Ruth Carr, a pilot cooking course including 'mothercraft', aimed at poorer village women, was examined. Like many other courses of this type, the objective was to train 'leaders' who would go back to the villages and instruct other women; however, most of the women had no interest in teaching anything, but had come to learn such skills as making 'proper wedding type cakes'. The course is seen as demonstrating that some basic assumptions about village women are incorrect, including the idea that poorer women are in touch with home-economics groups (which are in fact 'status groups', as the authors put it); women poor in resources feel inadequate to cope with these kinds of groups. They do not mention that poorer women would not have the spare time necessary to participate. Expatriate-led home economics courses, they decided, are unlikely to contribute much because there is no real basis for trust. One possibility would be to discard pretensions about helping poor women, and consciously meet the demand for prestigious home-economics skills from the women in the groups: 'We can take a realistic approach and make the élite women our target population and run courses in the skills that they ask for: e g cakemaking, icing, sewing machine use, flower arranging, hairdressing, fruit preservation etc.'[40] One suspects that similar criticisms could be made of other attempts to recruit and train rural women as 'leaders' of their communities, such as the courses in Kenya run by the FAO under the pretentious title 'Programs for Better Family Living'. The subjects taught, predictably, include home economics and related subjects like nutrition, child-care, 'home improvement', handicrafts and teaching skills.[41] The 'leaders' in many areas were very difficult to find; women attending in this capacity seemed to include young educated girls with no affiliation to any group, and even trained home-economics field staff.[42]

There is an inadequate scientific basis for the home economics projects and teaching which is evident in, for example, the nutrition and cookery teaching which is one of its most prominent aspects. Alan Berg, in his survey of nutrition as a factor in development, comments that most nutrition education is badly done. Medical schools ignore nutrition, reflecting perhaps its lack of prestige as a 'women's' or 'home economics' subject, as well as the long-established prejudice of the medical profession against preventive health care. Extension workers in nutrition in India — usually women — lack proper understanding of the program objectives. Girls training in the subject in home economics colleges rarely do any extension work. There is also virtually no evaluation of nutrition programs.[43] In general, Berg concludes, nutrition education seems to achieve little; the content is often 'intensely boring' as well as inappropriate to people's actual circumstances and diet.[44] There has been virtually no research on the nutritional value of locally available foods, particularly wild foods which are an important supplement to the diet. Instructors will recommend that villagers eat imported oranges for their vitamin C, completely ignoring local fruits which have far higher levels of the vitamin. Nothing is done to check changes in cropping patterns which have an adverse nutritional effect, such as the massive switch from sorghum and millet in Africa to maize and cassava. Deterioration in diet could even be promoted by nutrition 'education'. Surveys in north-eastern Tanzania, for example, have indicated that the increased consumption of prestigious cabbage has lowered vitamin A intake by reducing the intake of wild spinach.[45] Nutrition projects are open to the accusation that their main effect is to increase people's dependence on imported items, at great expense, with the overall effect of reducing both quality and quantity of the available food, a very serious issue for underfed people. John de Wilde states quite openly that women should be encouraged to purchase food for cash: 'given some knowledge of nutrition . . . the possibilities of spending more income on food and on types of food that may improve the diet are likely to be enhanced.'[46]

Another charge to which home economics projects are open is that of total irrelevance both to the needs of rural women, and even more to the resources available to them. For example, Uma Lele mentions that a home economics agent at LLDP in Malawi, who was trying to teach African women how to care for children, was observed washing a baby doll in an imported plastic bathtub filled with warm water. Neither hot water nor bathtub were available to the women. The agent also demonstrated the use of an imported disposable nappy — a luxury item that village women would never buy, and it would seem pernicious to suggest that any available income be spent on such non-essentials. Programs that attempt to impose additional tasks on women which are simply not feasible could even be considered as a theft of the women's time, a commodity which is perhaps the most crucial constraint to reaching an adequate level of family subsistence (see Chapter 7). Attempts to save women's time within the established home-economics framework, such as an elaborate time-and-motion study by the FAO on the arrangement of the stove, food-storage area and so on in very small houses, are observed by Uma Lele to be quite inappropriate, an application of techniques designed for high-income households which have little to offer in terms of Third World rural women.[47] They generally spend a very small proportion of their time on activities covered by home economics courses. They have a vast amount of expertise on local foods, cooking and childcare; they probably delegate much or all of the housework to children, particularly during peak agricultural seasons; and their biggest need is to conserve their own time and energy by reducing domestic tasks to a bare minimum. This is a strategy which home economics projects often seek to disrupt — and then their officials are distressed when their well-intentioned lessons seem to be ignored.

Like many of those involved in planning and administering development projects, home economists frequently have little respect for those they are ostensibly trying to help. The arrogance which Ehrenreich and English attribute to the early domestic science movement and its practitioners in the United States[48] can be a feature of work in the Third World. Shamima Islam describes rural women in Bangladesh in terms which show little awareness of the skills passed down from one generation of women to the next:

> '. . . women in rural areas of Bangladesh . . . are among the most tradition bound, conservative and are ill-prepared to assume the responsibilities for the development. They perpetuate a life style too inadequate for development and unfortunately, because of their socializing influence on children, they transfer to their children,

especially to their daughters, the same pattern of behaviour which seriously inhibits improvement in the rural areas.[49]

Handicrafts

A more sympathetic approach towards rural women, showing an understanding of their need to earn money for themselves and their dependants, is starting to emerge. It is becoming standard practice, in some agencies, to provide at least an element of income-earning activity in new projects for women. As Michael Moore, among others, has pointed out, the destruction of most handicraft work in the Third World by industrial products has displaced many workers, and whereas the men have usually had some alternative, however inadequate, women usually have not.[50] Ester Boserup underlines the enormous extent of the unemployment among women displaced from small-scale manufacturing.[51] At first glance, it might seem that the handicraft projects springing up for women all over the Third World offer the prospect of restoring some of the lost employment opportunities.

Unfortunately, close scrutiny does not bear out this optimistic assumption. The projects are almost all geared to a very limited and unreliable market, the tourist and overseas speciality markets. Objects produced are decorative, non-essential items. A typical project in the Dominican Republic, financed by the World Craft Council of the United States, deals in seven craft categories: ceramics, leather, doll and toymaking, textiles and fabrics, jewellery and associated products, wood and stone-carving.[52] Ester Boserup is very critical of this kind of approach, which may have some limited value in providing secluded women with a craft that can be done at home as 'the only possible first step towards bringing them into the labour market'. However, the effect for most women may be to drag them into low-productivity jobs rather than to help them find more productive and remunerative employment. She considers this kind of project as a form of compensation for discrimination in employment, and 'as a deliberate method of reducing the number of women competing with men' in the modern sector.[53] In terms of overall plans for promotion of rural industries, women have little more than token participation. In a very large survey of rural industries in Bangladesh financed by an AID contract, only one per cent of the field interviewers were to be women; Adrienne Germain concluded that this would mean women's industries being inadequately reported, if at all, with the result that they would be excluded from the design of new credit and industrial development programs.[54]

Germain makes a number of points about craft projects for women in Bangladesh which illustrate the inadequate foundations on which these are built. Despite the great emphasis on craft projects in that country, in almost no case has an analysis been done of how and why they succeed or fail, especially from an economic perspective. Nor has there been an attempt to measure the earnings of the women involved, whether this represents a fair return on their labour, and how it affects family welfare.[55] Since, in the largest and most successful of these projects, the Jute Works, it is official policy to hold wages down, below whatever is earned by husbands (if present),[56] it would appear that the profitability of this is to some degree at the expense of the poorest women producing the crafts. A very large proportion of these in fact are widows, and discriminatory pay policies would help to ensure that provision for their families remains inadequate. Ten thousand women are reported to be producing jute crafts for the Works.[57]

The whole question of selecting crafts suitable for women is extremely confused, but appears to reflect Western gender stereotypes regardless of the fact that the local division of labour, currently operating, is in direct contradiction to them. For example, a very popular form of craft project for women, in Bangladesh as elsewhere, is textiles and sewing. Considerable investments are required for the purchase of sewing machines and expensive raw materials from distant markets, to which women do not usually travel. Substantial skill in the maintenance and operation of the machine is also required which the women have to try to learn. Since commercial sewing is the virtual monopoly of men, who work in this field from adolescence, it is hardly surprising that sewing for a market beyond the confines of a woman's own family may be 'difficult, if not impossible', and that nowhere, with one exception, have any of the groups with sewing machines been observed to produce goods that can compete, even in the local market. Similar problems are apparent with weaving.[58]

Germain seriously questions the wisdom of trying to teach new craft skills to women, in crafts where local demand is nil; where they will probably not produce quality products for some time; and where raw materials are expensive and hard to obtain. Many of the 'production centres' running along these lines are often nothing more than sheltered workshops, and 'frequently women are not active participants in their organization'.[59] Another problem is the steady production which marketing, especially for export, demands. This is very difficult given the seasonal peaks in agricultural work and various other competing claims on the time and energies of women.[60]

In Botswana, craft projects for women are far less extensive than those in Bangladesh, but local representatives of various agencies are enthusiastic about the 'small industries' approach. The pattern of activities is equally stereotyped along Western lines: women are involved in machine-sewing and knitting; making jewellery with local stones; and leather-working. I observed the operation of Gaborone and Mochudi workshops and stores. In the sewing and knitting units, it was clear that investment in machinery and equipment was high; considerable skills were required, and many women had dropped out because of failure to learn the elaborate finishes and techniques. There was, in addition, pressure from the expatriate experts to reach the most sophisticated levels of production, rather than turning out mass-produced goods at a lower level, that would be available at a cheaper price to less affluent purchasers. One woman who was using only the most basic knitting equipment and producing an 'inferior' product, which her husband successfully sold from door to door in the new mining townships at very competitive prices, was in some danger of losing her place in the small industries block. The jewellery operation had started off very successfully, using local semi-precious stones; however, after a few years the sophisticated customers had become tired of the simple techniques used and more elaborate setting techniques were being discussed. With a falling-off in demand, it was becoming difficult to pay off the instalments on the polishing and other machinery. Although on one level a success, this operation supplies the local tourist and South African market with the employment of only two or three women. Leatherwork is also a reasonable success, but demand for the products is not very good; they are neither sufficiently cheap nor sufficiently elaborate to sell in large quantities to tourists, or in the export market.[61] Prospects for expansion are also rather unfavourable. In general, the operations are capital-intensive, employing few women; the market is restricted and many women fail to complete the courses. These operations are, on the whole, unrelated to demand for essential items by poorer people in the country itself. They certainly offer no solution to the needs of poor women to increase their agricultural productivity and/or their cash income.

The prospects for Third World crafts, as currently being

developed by well-meaning development organizations and 'alternative marketing organizations', have been carefully evaluated by two specialists in the area, Jacqui Starkey and Maryanne Dulansey. They point out that any viable system would require matching products with the market; improving feedback between market and producers, including the forecasting of market trends; and improving the organizational, managerial and business skills of producer groups and the application of quality controls. These elements are, so far, woefully lacking in the rather amateurish efforts of so many organizations in this field, which effectively retards the development of the whole craft sector. In addition, the reliance on tourists, the luxury market and exports is badly misplaced; Starkey and Dulansey point to the need to re-evaluate the indigenous market, whether local, national or regional, as part of a strategy aimed at self-reliance.[62] At present, they argue:

'. . . Third World craft producers, especially the poorer of them, labour under constraints which limit their ability to gain a meaningfully large share of the world market . . . Alternative marketing organizations are in a position to mask this reality, and to create expectations and conditions of dependency which, though unintended, have a counter-developmental impact.'[63]

Special projects for women: help or hindrance?

As already mentioned, evaluation is virtually non-existent for most women's projects, whether in relation to home economics, handicrafts or some of the other approaches now being developed. The fact that rural women do not respond very enthusiastically to home economics projects, particularly in comparison with other kinds of projects, is occasionally recognized. A report on the UNDP projects to provide an 'adviser on women's interests' (defined as nutrition, child-rearing and the rest) in Cameroon stated that it 'has failed to create a real impact on both the situation of women and the contribution they are able to make in economic activities'.[64] A training program in home economics in Tunisia 'was not successful'.[65] Many of these unsuccessful projects are quietly terminated, and in common with failed development projects generally, there is a marked reluctance on the part of the agency involved to discuss what went wrong.

An attempt has been made to compare home economics projects with others, in terms of women's response to them, under an AID contract with Development Alternatives Inc.

Donald Mickelwait *et al*, while making great efforts to find good qualities in home economics projects in the seven Third World countries surveyed, concludes:

> 'Home economics activities may provide women with training that is useful for them . . . However, as promoters of change, modernization and development, such courses rank far below other types of activities. In several African projects . . . it has been found that major behavioural changes by women came about significantly faster through activities dealing with agricultural production than through family care [sic] projects.'[66]

In Bolivia, where projects stress such allegedly 'traditional' skills as sewing, cooking and flower arranging, the courses were often found to be incompatible with women's productive work. Where women and girls do take up the training, it is seen as a factor in their heightened migration to the cities to take domestic employment — hardly a benefit to the rural sector for which the project is ostensibly designed.[67] What this seems to indicate is that either a home economics project is irrelevant, and therefore has negligible impact; or, if it does have an effect, this can be economically and socially disruptive.

Any volunteers?

One common and unfortunate element of women's projects, which is not confined to home economics, is the assumption that women can be trained as 'leaders' who will work for nothing. Consternation is expressed by the planners when this turns out not to be the case with poor women, to whom working time is of great value. In Botswana, expatriate officials, led by a very highly-paid 'volunteer', have organized functional literacy groups for the Botswana Extension College. They were outraged when the leaders, 90 per cent of them women who were being paid two Pula (about £1) a week for six hours of teaching in very difficult conditions, left to take up school-teaching jobs. The original idea had been to pay nothing at all, on the grounds that volunteers would be better motivated. One of the teachers leaving the scheme objected, 'You are killing us' and this seemed to express the feelings even of those with no alternative to staying — but the feelings of the Batswana women seemed to have no effect on the plans of the American 'volunteer' man in charge.[68]

The official expectations about women's motivation can be hopelessly unrealistic. Discussions about a new approach to women's projects in the Upper Volta AID office, for example, tended to assume that village women working for them would not be paid; allegedly the project could not afford it (although it could afford to pay $30,000 for three American 'experts').[69] It was argued that the job would 'develop a status' for the women as compensation for not being paid; or alternatively, that the activities of the village groups might generate a cash profit which could be paid to the women working on it.[70]

In Botswana, the impact of the Mochudi Homecraft Centre is greatly diminished by the fact that very few of the school-teachers are prepared to attend the home economics courses. This is because they receive no salary while on the course, have no automatic rights to an increase in salary for the skills acquired, or even a guarantee of re-employment by their District Council at the end of the course.[71] The neatest solution to the problem of paying màle extension workers and not women is perhaps that adopted in Bolivia, where the unpaid women were so much more successful that a program of partial salary payments for the men living in rural villages was discontinued. The reaction of the men involved to the introduction of this 'women's project' is not recorded.[72]

The expectation that women will do important work in the community for nothing is very much a Western, upper-class idea. Formal voluntary work on any scale is not done by poor women in Western countries, let alone in Third World rural areas. John de Wilde mentions the inadvisability of using volunteers as extension workers.[73] In addition to the impracticality of any program based on the donation of precious time by poor women, this approach is necessarily related to the amateurish and patronizing (or maternalistic, as it has been dubbed) approach of project organizers. Mickelwait *et al* acknowledge that a by-product of encouraging voluntary work in official women's programs, which they consider desirable, is an increase in the involvement of female dependants of the male élite. As they candidly admit, the National Councils of Women to be found in so many countries 'are often made up of the female relatives of government officials'.[74]

The alibi effect

As long as 'women's projects' are set up in apparent competition to the mainstream of economic development projects, they will remain fringe operations with little prospect of making any real impact. Home economists in the FAO are well aware of the problem as it relates to their projects: women's issues are low priority in terms of funds as well as prestige and the first to be cut if the agency involved hits a financial crisis.[75] As 'social development' or 'social welfare', women's projects have been declining in real terms over the years in which economic development has been seen increasingly as the only issue worth serious consideration. Discussion about women's projects, and to some degree the introduction of new projects since 1975, is a precarious achievement which depends on continued agitation by the minority of concerned women in the development organization involved: a case of running exceedingly hard to stay in the same place.

An example of the problem is the women's project run by the UN Office of Technical Co-operation in Swaziland, a new departure using labour-saving technology for various basic tasks in order to free women for income earning activities and agricultural production. The women have been very enthusiastic and have shown a willingness to experiment with the new equipment offered. While on paper and in public the Swazi Government has been equally enthusiastic, in practice they have managed to obstruct and frustrate its operation. According to Africans involved in the program, all imported items and new staff appointments require official authorization, which is somehow never forthcoming even for the most minor items. They have concluded that the enthusiastic rhetoric by the King and his officials is a cover for attacks on a project which would significantly benefit women. The project makes an easy target, since it is conspicuous and is unrelated to any other development program; an integrated project could never be singled out in this way for official obstruction.[76]

The vulnerability of a women-only project is not limited to attacks by the government. A report on community development in Tanzania states:

> 'It was considered important not to isolate the women too much for the purpose of learning new skills, and so create the possible impression of imparting to them an exclusive mystique. Otherwise, as past experience in rural areas had shown, husbands sometimes grew suspicious . . .'[77]

Such hostility can easily eliminate a women's project. Women's hostility to a project for men only is unlikely to have the same effect except by more subtle means of subversion, usually not recognized as such by project officials. The possibility of men's hostility to projects for women only was recognized as so serious a problem for the big Upper Volta project, backed first by

UNESCO and now by UNDP — the Project for Access of Women and Girls to Education — that it virtually ceased to be aimed at women and became one of the few truly integrated projects open to women and men alike.[78]

Ironically, despite this high degree of integration the Upper Volta project is used extensively as an alibi for the segregation of projects, to make virtually all of them for men only. The enormous, multi-donor Onchocerciasis Control Program (OCP), covering several countries of the region and based at Ouagadougou, ignores the existence of women in the area. When asked how women were involved in their various programs, and particularly in planning for resettlement of the treated areas, the officials reacted defensively almost to a man, saying that to discuss women one should go to the local women's project.[79] Also in Upper Volta, the local AID representative answered my questions about the Congressional mandate to integrate women in development by listing a series of special surveys and small projects, most of them not yet in operation. When asked how these special projects related to integration of the mainstream projects and programs, he showed considerable embarrassment and terminated the interview without answering the question.[80]

Adrienne Germain describes a similar situation in Bangladesh: donor attention to women's needs tends to be segregated from their major development programs, and, as a direct result of this segregation policy, large numbers of women may be excluded from broader development programs which could adversely affect the women's interests.[81]

Special segregated projects for women not only provide an excuse for failure to integrate all development projects; they are also ammunition for counter-attacks by male officials. Several of those I encountered in the field offices of international development agencies commented caustically on the fact that women were asking too much: equal access to the existing projects as well as additional projects for women only. Several of them suggested, with some hostility, that women were actually getting a better deal than men. Generally speaking, they seemed to be using the special projects to attack the demand for equal access to major projects, rather than the other way round.

In theory, the men have a case. A UNDP paper on integrating women into development cites the Regional Bureau for Africa as suggesting that the UNDP offices concerned, together with governments, study the feasibility of designing special programs for women — 'in addition to possibilities of incorporating into on-going projects activities to enhance the role of women in development'.[82] As long as a few marginal 'women's projects' are being put forward as a token expression of commitment to the integration of women in development, any movement towards actual integration is likely to meet with strong opposition from the men in control. It would be quite feasible, in fact necessary, to use the resources now devoted to the 'women's projects', including much of the home economics activity, as the basis for integrating development projects. In many cases, the special projects are virtually monopolizing the best of the trained women available who would provide a valuable nucleus of women for the major projects. The longer the delay in integration the greater the momentum accumulated in the construction of a women's ghetto of marginal projects. This will inevitably result in the creation of a vested interest struggling to maintain the women's sector rather than integrating the mainstream. Increasing numbers of women and girls will be channelled into home economics or other very limited and stereotyped fields (secretarial, fashion, etc) as they enter various levels of formal and informal education in larger numbers. There is a place for work on such topics as nutrition, promotion of handicrafts, child health and the application of appropriate technologies to cooking, water collection and other non-field tasks, currently dealt with under the home economics label. These need to be developed and applied in a professional manner. It is the relegation of such activities to the stereotyped and confused heading of 'home economics', as if they were of exclusive concern to women, that has led to their being, in so many projects, part of a component which is second-best.

As long as women are restricted to training in women's projects, in specifically chosen women's subjects, their opportunities will be severely limited by the lack of resources for these. At LLDP in Malawi, for example, even the 'women's training' sector, to which women are consigned, is not reaching the planned levels, because it depends on the output from the home economics course at the Tuchila Farm Institute and, in future, will depend on the same course at the Natural Resources College to be established at Lilongwe.[83] Uma Lele notes that staff employed in LLDP's community development work, who are responsible to the district council rather than LLDP, 'reflect the generally poor investment in women's trainers'. She cites Bill Kinsey to the effect that their training is inadequate and their salaries low; hence, there is a very high rate of turnover.[84] This loss of home economics staff seems to be a common feature of many rural development programs and an indication both of the inadequate training, field support and salaries paid to the women. There is also a great waste of resources involved in the relegation of some of the most able women and girls available to a second-class sector.

Tapping the real motives

Elimination of much of this waste would seem to depend on a recognition by planners that women are motivated by much the same interests as men: tangible benefit from the project in question. The UN report on women in Botswana points out that functional literacy teaching 'must be functional in a personal beneficiary way [sic], often interpreted by the women as personal economic gain'. They also feel the need to learn how to operate in a new social structure, new legislation, their own civil rights and so on.[85] This would require a different approach to the current home-economics one. In Upper Volta, Joy Greenidge found that women wanted help mainly in growing cash crops (in communal fields), making clothes for sale, and in the establishment of a store as an income-producing venture.[86] All these activities were turned down because of the lack of money for 'women's affairs'.

In Bangladesh, in contrast to what is expected of Islamic societies, Adrienne Germain reports that women participate in programs outside their homes when they are offered, and that economic gains may be the major motivating factor. Some women persist despite substantial male opposition.[87] The 'Women for Women' group in Bangladesh argues strongly that training should aim to make women self-sufficient, because so many of them face destitution at some point in their lives, and not aim to make them merely 'more skilful dependents' [sic].[88] A study of rural women's groups in Colombia, Korea and the Philippines concludes:

> 'The most successful groups observed to date all make financial gain to the family a high priority. One result . . . appears to be easier acceptance of the groups by men in the community. Another is that the status of rural women increases with their value as wage earners.'

So much for the idea that 'status' will compensate for lack of income. The study stresses that nothing compares with 'the perceived economic gain of the accumulation of capital or cash.'[89]

Mobilizing women's groups

Although it would seem inadvisable, for the reasons stated above, to confine women to special projects away from the mainstream, it is important to bear in mind the significant contribution which the existing women's groups of various kinds can make to development. In many areas, women have always worked communally, or had their own groups for activities such as revolving credit. Women are a strong force behind many self-help schemes, such as *harambee* in Kenya — a concept different to that of Western-type voluntary work, in that large numbers of women co-operate in some activity, such as building a community school or clinic, from which each expects to derive a tangible benefit. Sometimes the more formally organized women's groups do important work which deserves support from development agencies. In Botswana many of the local groups belonging to the two major Botswana organizations, the Botswana Council of Women and the YWCA, as well as church women's groups, have built classrooms, clinics, latrines and bus shelters. They have also pioneered the provision of day-care centres, for which they now have to bear the running costs and provide voluntary staff (a major constraint on expansion of this much-needed service). Four women trained at great expense in Israel as nursery-school teachers for the BCW are unemployed because of a shortage of funds for their salaries.[90] The integration of such efforts into bigger community development or education programs could lead to their receiving funds with which to expand these activities for the particular benefit of the large numbers of single mothers and older children, particularly girls, whose child-care duties conflict with regular attendance at work or school.

Home economics employees, and the groups of women they initiate, can be retrained to make their activities more relevant to the needs of rural areas. For example, home extension workers in Zambia, trained in home economics and in organizing women's groups as well as conducting courses for women at the farmers' training centres, told me that by far the most popular topic among village women was agriculture; about three-quarters of the time was spent on this. The 'home extension' workers were now spending much of their time distributing seeds, helping with agricultural techniques, and demonstrating food preservation by sun-drying. When they had first started the groups, the women had expected to be taught sewing; they changed their attitude radically as food became scarcer and more expensive, and became keen on agricultural techniques which they had previously dismissed as dirty, and for men only. The extension workers themselves were very well motivated, and although young and mostly unmarried, were determined to continue working regardless of whether or not they married, or what their future husbands might think on the subject. They cited the rising cost of living as the main reason for wanting to continue, and implied that they valued their economic independence. The motivation found here was probably related to the fact that salaries were reasonable, and also that they were about to start training courses in horticulture, at the training centre which was headed by a woman with whom they identified strongly.[91] In effect, if not in name, they were becoming agricultural extension officers.

'Home extension' groups in the surrounding area demonstrated the same phenomenon of a shift in demand from sewing to agriculture. One striking feature was the expenditure of time necessary to provide even minimal funds for the groups; the traditional beer-brewing was one means of doing this, another was applying the rather irrelevant sewing and embroidery techniques to making tableclothes and napkins which they hoped to sell (obviously not to most people in the village). This seems to be a very common feature of women's village groups: they do a lot of work to raise funds, which, although valuable as an indicator of the motivation and group cohesiveness available, restricts time and energy for more directly productive activities. One of the leading women in a local group impressed on me the urgent need for some money to pay for the training and equipment the women needed (one example being village wells): 'Go and tell the big men we need money. We've done everything we can to get it for ourselves. They've got plenty, why don't they give us women some?' The village was very close to the horticultural training scheme.[92] A similar situation was to be found in Botswana, where a primary school teacher had started some women's groups on a purely voluntary basis. It was outside the confines of a big scheme funded by external organizations which trained boys in vocational skills under the 'brigade' system and she had no access to any development funds at all. The women in the group made bread and brewed beer to raise some money for their activities.[93]

A number of women's activities seem to develop quite spontaneously in various places; Adrienne Germain mentions women's co-operatives in Natore, Bangladesh.[94] In Ivory Coast, I was told that women had formed labour co-operatives for working on each other's fields and to hire themselves out to others for a fee; they received no support from the well-funded co-operative development program.[95] Many of the new groups springing up, as well as the more traditional communal activities based in a village or especially among secluded women in an extended family, could be important channels for development funds from the major development projects, if women were no longer seen as a separate and marginal sector. In Botswana, the Co-operative Development Program, considered by the sponsoring organisation, ILO, as the most successful in Africa, seems to owe much of its success to the participation by women. This is particularly true in consumer and credit co-operatives, because of the women's strong tradition of similar forms of co-operation at village level. Perhaps 90 per cent of the co-operative members are women, and they provide a major proportion of the officers. The official training courses, however, are attended almost entirely by men.[96]

Women can be an important resource for development, and women's groups an effective channel for funds aimed at meeting the needs of the poorest people in rural areas of the Third World. Their potential can best be realized if they are integrated into the whole spectrum of development programs, and not relegated to the marginal sector currently reserved for women.

A note on 'population' projects

The one area where women are not treated as marginal is in the proliferation of population projects. But here one finds much of the wastage of resources that is evident in the marginal 'women's projects', although on a much greater scale.

Population programs are planned and organized mainly by men, and aimed almost entirely at women — women as objects whose fertility is to be controlled (hence the phrase 'population control') rather than as people who would wish to control their own fertility. It is assumed that women do not know what is good for them, and that they must be persuaded to become 'family planning acceptors'. Targets are set; the negative effects of various kinds of contraceptives played down; and very often the choice of contraceptives to be made available is set by the planners, based on a cost-benefit analysis. In many countries, a few methods might be free while others carry a charge. Coercion is used, or is frequently suspected, and people's sensitivities about such issues as sterilization without popular consent — an explosive issue particularly where there is racial tension — are often ignored.

It is becoming clear that the most ambitious case so far of an attempt at 'population control' — in India during the Emergency — has been catastrophic for family planning, even in terms of the targets set up as the criteria for such 'control' projects. Following the programs of forced sterilization of both women and men, there has been a steep drop in voluntary sterilizations in the states affected — while several of the southern states which were largely unaffected by coercion continue to show a rise in sterilizations, and an overall fall in the birthrate. The 'Sanjay factor' has led to a serious demographic setback, with all population targets undermined by several years; the International Planned Parenthood Federation (IPPF) expects that family planning will take years to recover. In terms of the failure of the present government to restore confidence after the end of the Emergency, it is estimated that this alone has resulted in three million extra births, and the death of 16,000 women in childbirth.[97]

Another example of the counterproductive nature of a 'population control' program is Tunisia. The program was started in response to rapid population growth. In the early years, 1964-66, the pressure to recruit large numbers of volunteers led in some cases to a lack of adequate attention to follow-up care for women who had already accepted. This is particularly important to the women most in need of contraception because of malnourishment, anaemia and poor health generally.[98] In Morocco, the contraceptive pill was seen by poor women as feasible only for richer women with a good enough diet to eliminate the severe side-effects that they experienced from it.[99] The IUD has gained a particularly bad reputation. A report on the Tunisian program by SIDA, the Swedish aid agency, observes: 'Had the initial goal of the program been women's health and had its structure been different, the result would surely have been better.'[100]

The inadequacies of all available contraceptives need not in themselves be a barrier to acceptance; the problems seem to arise out of attempts to force people, as in India, or to mislead them by failing to explain the possible side-effects or to offer treatment when complications arise, as in the Tunisian case. The pioneering birth control clinics in Britain, offering relief to working class women with large families in the Depression, were 'all the more effective because they made no attempt to gloss over the shortcomings in the techniques then available,' according to Peter Fryer.[101] A 1932 report on clinics in the Midlands concludes that their success in the worst-hit areas, including those set up during a lock-out of miners near Wolverhampton which led to starvation, 'refuted the assertion . . . that miners' wives did not want birth control advice' and proved that the need was even greater in remote areas than in large towns.[102] It was found particularly helpful to employ only women as medical officers, and to offer a secluded clinic location that enabled poor women to visit in private.[103] Somewhat belatedly, the same lessons are being learned in some of the family planning programs in the Third World. In South Korea, Nepal, Thailand, Indonesia and other countries, women's organizations or local midwives have been found to be very successful channels for disseminating information, support and suitable contraceptives to village women. Particularly striking is the high continuation rates, a very weak point in many programs.[104] In South Korea, for example, many of the women were found to be willing to accept IUD insertion provided it was done by a woman, and at home. They almost invariably preferred the village midwife to the male doctor.[105] In Thailand, when midwives were trained to prescribe pills and insert IUDs, with assistants doing home visits, the number of new acceptors tripled in one year.[106] An interesting and perhaps representative list of conditions for acceptance of family planning was drawn up by a meeting of women in a Central Java village in Indonesia; they would gladly accept provided it:

'— not interfere with our working;

'— not do us permanent harm;

'— not be against our religion;

'— be free or nearly so;

'— have a woman to examine us and to teach us what to do and how to do it;

'— remain a secret between her and us.'[107]

Dealing only with other women allows women control over their information and supply networks, free from the fear of comment from men and boys who could be expected to pass moral judgements about women's actions without themselves facing any of the dangers and tragedies of pregnancy and childbirth for both women and newborn children. These are among the most powerful motivations for women in adopting family planning. In many areas, men are not even expected to maintain their own children and it is, in any case, much easier for fathers than for mothers to evade their responsibilities towards children.

The lengths to which women will go to control their own fertility are evident in the very high incidence of illegal abortion, in countries with few or no contraceptive facilities and a prohibition on legal termination of pregnancy; this in spite of the serious risks to life and health involved in illegal abortions. Because of 19th century anti-abortion legislation, a prime example of Western male ideology imposed throughout the world by means of colonial policy, the topic has until recently been a forbidden one, and data scarce and unreliable. It is becoming increasingly evident, however, that abortion is the major means of fertility control in the Third World today. One estimate, for example, suggests that half of all pregnancies in Brazil end in abortion. Estimates of the total annual number of abortions worldwide range up to 55 million.[108] Where abortions are carried out clandestinely, in unhygienic conditions, they result in enormous numbers of deaths and injuries.[109] In some countries, abortion is the main cause of death among women of child-bearing age. In Chile it is estimated that complications from illegal abortion, even though only about one-third of the victims obtain medical help, account for 25 per cent of blood bank resources, 35 per cent of hospital beds, and a third of all operations – in a country desperately short of medical facilities.[110] This appears to be fairly standard for many Third World countries. In Africa, urban women are estimated by an IPPF study to have a 40 per cent chance of admission to hospital at some time for treatment of complications from abortion (either spontaneous or induced).[111]

Abortion legislation is now being liberalized in many countries, although mainly in richer countries so far.[112] In India, one of the few Third World countries where abortion was made legal for most women in 1972, Baskara Rao and Ramesh Kanbargi report that, since legislation, there has been a decline in the age of women obtaining abortions, their family size, duration of pregnancy, and age of new contraceptors after abortion. The proportion of women with less than primary education using contraception has increased, as has the number of new contraceptors following abortion. This indicates rapidly growing acceptance of the government's program for abortion, which is relatively safe, and the positive effect this has on contraception. There is considerable scope for improvement in the service; Rao and Kanbargi conclude that 'if there were more approved rural hospitals and a greater diffusion of information, the number of legal abortions would increase further.'[113] Safe abortion could be described as one of women's greatest unmet needs.

It is commonly assumed that Third World populations are rising because of what one writer terms 'une natalité du Moyen Age en face d'une mortalité moderne'.[114] However, it is by no means true that Third World women have always borne large numbers of children. Evelyn Reed cites 19th century European travellers, among native Americans of both North and South America, as finding that they had far fewer children than contemporary European women and with a lower mortality rate. She points out that one means of control, imposed by a group of women in a polygamous marriage, was prolonged periods of segregation of wife and husband each time a child was born; and that this is not possible under 'modern' or 'developed' patterns of male-dominant, monogamous marriage with a common residence.[115] C M N White, writing about the Luvale of Zambia in 1959, found that their fertility ratio was extremely low, and the net reproduction rate was just at replacement level[116] — a little-known case of Zero Population Growth (ZPG). Similar phenomena have been found in gathering and hunting groups, which have been the subject of particular interest by Europeans.[117] Ernestine Friedl points out that the work of gathering, performed most commonly by women, requires a low fertility rate. Women must be able to walk long distances without excessive burdens.[118]

Colonial policy, it seems, was strongly pro-natalist, reflecting both the male ideology of contemporary Europe and also a special interest in acquiring large numbers of labourers for commercial and other colonial ventures. The introduction of anti-abortion legislation, already mentioned, was part of the same approach, heavily overladen with moralistic overtones. Missionaries campaigned vigorously against a number of measures by which women controlled their fertility, including the segregation of the sexes after childbirth, the whole practice of polygamy, and women's secret societies used for, among other things, initiation in traditional methods of fertility control.[119] Mary Kingsley described a missionary in West Africa at the end of the 19th century promoting bottle-feeding as a means of encouraging sexual relations between married couples.[120]

There is, at present, no adequate data from which to assess the influence of colonial policy on women's loss of control over their own fertility. What is clear, however, is that population growth rates in almost all Third World countries are extremely high, and that the only feasible means of reducing this is by birth control, which depends largely on the motivation of women. In fact, despite the family planning programs that have been under way for a relatively short period of time, and despite their innumerable deficiencies, Third World women are already using them to limit their own fertility to an extent which is having a significant effect on the birth rate. A comprehensive analysis by the Office of Population in the US Agency for International Development (AID) concludes that between 1965-1974 the world birth rate declined from 34 to 28.2 million live births per thousand, and the population growth rate declined from 2 per cent to 1.63 per cent. The turning-point came in 1970-71, and an accelerating downward trend is now 'firmly established', according to the Director of the Office of Population. He has suggested that the most notable discovery of recent years is that poor and illiterate peasants control their fertility to the same extent as literate urban residents, when the means are fully available.[121] Fertility decline is not limited to small, often urbanized countries and islands; it has also been a factor in some of the most populous countries, with more than a 20 per cent decline in three of the 13 Third World countries with populations over 35 million, and declines of between 10 and 19 per cent in another six, between 1965 and 1975.[122]

The whole question of the control of fertility is complex, and the subject of extensive literature. One of the most interesting recent trends in this literature has been the attempt to correlate fertility with the 'status of women'. Reduced fertility can often be correlated with 'status' as measured by such factors as education, employment, legal rights, characteristics of the family and so on.[123] The problem is that the concept of 'status', as already noted, is an extremely nebulous one. If it is assumed, for example, that the level of education and income is an index of 'status', then apparently inexplicable anomalies arise such as the phenomenon, in many countries, of high fertility at the upper end of the education and income scale. The correlation overall is uncertain, and often fluctuating for no apparent reason.[124] The difficulty is resolved if the obscure concept of 'status' is discarded in favour of the more specific one of control. Whereas 'status' refers to an unvarying condition of women, reflected in some indexes such as that of education or employment, control is divisible in terms of the specific activity or resource at issue. Thus, women may or may not have access to adequate facilities for contraception; they may or may not control sufficient cash to purchase contraceptives, particularly those which would be most congenial to them; they may or may not have the necessary information about what to ask for.

Among other variables is the extent to which they control their own children, in terms of custody, use of their work and continuing contact in old age; the influence women have on decisions within the family unit, if any; and their access to activities other than child-bearing with which to justify their existence. Also important is the likely contribution of sons and daughters respectively to their parents' income; preference for sons is a major influence in family size. Although often connected, these elements are quite separate and cannot be reduced to a single 'status', particularly as measured by some more or less arbitrary factor such as education (which could appear relevant only by reflecting female access to cash for school fees, for example).

It has lately been recognized within UNFPA and a few other agencies that the real problem is not one of motivating women to control their own fertility but the hostility of the men towards the use of contraceptives. This is frequently formulated in terms of the man's loss of control over their wives' sexuality.[125] The conclusion has tended to be that population propaganda should be directed at changing men's attitudes — an uncertain and very lengthy process at best, since men incur no particular costs in having large numbers of children, especially in rural areas. It would be much more effective, in fact, to recognize that women will control their own fertility if they have the means such as cash, information and local networks to do so.

It is far too often assumed in family planning programs that all women form part of a Western-style 'couple', and that fertility decisions are discussed between husband and wife. In the first place, many women, perhaps the majority in some areas, do not form part of a 'couple'; and discussion is in fact more likely to take place between women themselves, including co-wives, than between wives and husbands, even where a couple can be identified. Judith Bruce cites evidence that many married women in the Third World use contraception without informing or consulting their husbands.[126] This is of course a feature of most, if not all traditional methods of fertility control, particularly menstrual regulation and abortion. Ruby Rohrlich-Leavitt et al, for example, refer to the frequent abortions of Australian aborigine women without reference to men.[127] In Egypt and Mexico, also, women have been found to prefer contraceptive techniques which closely parallelled their own traditional methods; in Egypt, a major reason given was that this did not require visits to a clinic, and the husbands did not have to be told.[128] Formal religious codes forbidding certain kinds of contraception or abortion, which planners so easily assume to be a powerful deterrent to fertility control, may be completely ignored by the women involved. Fatima Mernissi's important study of family planning practice in an urban slum area in Morocco shows that women 'are generally not easily impressed or influenced by formal religious or legal codes. Family planning officials apparently make a mistake if they ascribe too much importance to religious and legal authorities in the formation of women's attitudes toward fertility control.' The formal codes would probably influence the husbands; the women organize their own efforts at fertility control among themselves without reference to their men.[129]

Many of the emerging attempts to reduce the birth rate by 'raising the status of women' are useful because they help to meet the need for women's control over their own fertility. For example, projects which enable rural women to earn some cash have been found to contribute to acceptance of family planning. All kinds of complicated indirect effects are claimed for this, which include the inevitable 'status-raising' but also the improvement in nutrition and health, a reduction in the defeatism characteristic of the extremely poor, and a reduced need to have children as social security.[130] A far more direct effect, however, would be the availability of cash for contraceptive devices and perhaps travel to a clinic, as well as the more indirect benefit of the confidence arising from the increased control of resources generally, which cash invariably involves. Family planning projects would be more effective if they defined their objectives in terms of women's control over specific resources related to contraception; and on a wider scale, control over their whole lives. Since it is increasingly the women who are responsible for maintaining children, such an approach would help to ensure that the children born have a better chance of survival.

Richard Titmuss makes the intriguing suggestion that the increasing control by British women over their own fertility has been the most crucial factor in raising working-class levels of living since the industrial revolution.[131] The rapid decline in family size for all social classes after 1900 was prompted by a massive and little-studied women's movement for control over sexuality (especially through the widespread adoption of the withdrawal method).[132] This, suggests Titmuss, meant:

'. . . in terms of family economics, a rise in the standard-of-living of women which has probably been of more importance, by itself, than any change since 1900 in real earnings by manual workers.'[133]

Perhaps family planning programs, if reorientated towards women's needs, could have similar impact on their standard of living in the Third World today.

References

1. A British Ministry of Overseas Development official has informed me that this is the current question there in response to International Women's Year and associated agitation about women.
2. International Labour Office, *Equality of opportunity and treatment for women workers*, Geneva, ILO, 1974, pp 72-76.
3. Kathleen Newland, *Women in politics: a global review*, Washington DC, Worldwatch Institute, Worldwatch Paper 3, 1975, p 10.
4. Adrienne Germain, *Women in development: final consultancy report, March 8 — April 20, 1977*, Ford Foundation internal communication, 25 April 1977, pp 2-3.
5. *Ibid*, p 3.
6. Cited in Adrienne Germain, *Consideration of women in the two-year plan 1978-80*, Ford Foundation internal communication, 20 June 1978, p 1.
7. *Ibid*, p 3.
8. Adrienne Germain, 1977, *op cit*, p 4.
9. Thanhdam Truong, *Report on current UNDP-assisted projects involving women*, prepared for the UNDP Division of Information, mimeo, October 1974.
10. *Report on development assistance to Malawi, 1976, by the Resident Representative of the United Nations Development Program in Malawi*, mimeo, April 1977.
11. Thanhdam Truong, *op cit*, p 9.
12. Ulla Olin, 'A case for women as co-managers: the family as a general model of human social organization', in Irene Tinker, Michele Bo Bramsen and Mayra Buvinic (eds), *Women and world development: with an annotated bibliography*, New York, Praeger, 1976, pp 105-28.
13. See UNDP, *United Nations Development Program in the Republic of Upper Volta: assistance to activities related to the implementation of national, regional and world plans of action for the integration of women in development*, paper for the African Regional Conference organized by the Economic Commission for Africa, Nouakchott, September 27 — October 2, 1977.
14. Interviews by the author with officials of the UNDP Resident Representative's office, Ouagadougou, October 1977.
15. UNDP project document, *African Training and Research Centre for Women, and African Women's Development Task Force (RAF/75/036/A/01/01)*, August 1975.
16. UNDP, *Activities related to the implementation of national, regional, and world plans of action for the integration of women in development*, paper for the Nouakchott Conference, 1977, p 1.
17. UN Office of Technical Co-operation, *United Nations Assistance in*

women's activities, Botswana: based on the work of Karen Poulsen, UN document ESA/OTC/REP/75/1, 1975, p 29.

18. Brenda Gael McSweeney, *Regional conference on the implementation of the national, regional, and world plans of action for the integration of women in development,* Report to UNDP, mimeo, October 1977, p 18.

19. FAO, *The family in integrated rural development,* Rome, FAO, 1975, p 51.

20. Natalie Hahn, *Brief for FAO representatives — women in agricultural and rural development,* mimeo, 25 July 1977, p 2. See also FAO, *An inventory of FAO activities as related to women,* the FAO Inter-Division Working Group (IDWG) on Women in Development, February 1977.

21. See Barbara Ehrenreich and Deirdre English, 'The manufacture of housework', in *Capitalism and the family,* San Francisco, Agenda Publishing Co, 1976, pp 7-42.

22. *Ibid,* p 31.

23. Adrienne Germain, *Women's roles in Bangladesh development: a program assessment,* Dacca, Ford Foundation, 1976, p 6.

24. *Ibid.*

25. *Ibid,* Appendix A, p 7.

26. Ester Boserup, *Woman's role in economic development,* London, George Allen and Unwin, 1970, p 220.

27. *Ibid,* p 222.

28. UN Economic Commission for Africa, *The role of women in African development,* UN World Conference of the International Women's Year, UN document E/CONF. 66/BP/8, 1975, pp 18-20.

29. Gloria Scott, *Integration of women in development: the impact of some development projects,* unpublished working paper, 1976, pp 11-12.

30. Uma Lele, *The design of rural development: lessons from Africa,* World Bank Research Publication, Baltimore, The Johns Hopkins University Press, 1975, p 77.

31. Lilongwe Land Development Program figures.

32. Uma Lele, *op cit,* p 167.

33. Adrienne Germain, *Report on consultancy, women's programing, June 1-23, 1978,* Ford Foundation internal communication, 28 June 1978, p 5.

34. John de Wilde, *Experiences with agricultural development in tropical Africa,* Vol I: The Synthesis (published for the International Bank for Reconstruction and Development), Baltimore, The Johns Hopkins University Press, 1967, pp 191-92.

35. Interview by the author with Carolyn Barnes, Project Director, Ouagadougou, October 1977.

36. Interview by the author with Joy Greenidge, Ouagadougou, October 1977.

37. K G Rama, *Women's welfare in Tamil Nadu* (published for the Madras Institute of Development Studies), Madras, Sangam Publishers, 1974, p iv.

38. Lucy Mair, *Marriage,* Harmondsworth, Penguin, 1971, pp 16-17.

39. Adrienne Germain, *op cit,* appendix A p 4.

40. Marit Kromberg and Ruth Carr, *Report of the 'pilot' VHAL cooking*

course at Mahalapye, October 6-8th 1976, mimeo, 1976.

41. Food and Agriculture Organization, *Women's leadership in rural development: report on a national workshop to co-ordinate and plan for the women's group program* (Programs for Better Family Living, Report No 14), Nairobi, FAO, 1975, p 2.

42. *Ibid*.

43. Alan Berg, *The nutrition factor: its role in national development*, Washington DC, The Brookings Institution, 1973, pp 78-79, 83.

44. *Ibid*, pp 85-86.

45. J Kreysler and C Schlage, 'The nutrition situation in the Pangani Basin', in H Kraut and H D Cremer (eds), *Investigations into health and nutrition in East Africa* (Afrika-Studien No 42), München, Weltforum Verlag, 1969, p 170.

46. John de Wilde, *op cit*, p 211.

47. Uma Lele, *op cit*, p 119n.

48. Barbara Ehrenreich and Deirdre English, *op cit*, p 30.

49. Shamima Islam, *Women, education and development in Bangladesh: a few reflections*, paper for the seminar on 'Role of women in socio-economic development in Bangladesh', Bangladesh Economic Association, Dacca, mimeo, 9 May 1976, p 8.

50. M P Moore, *Some economic aspects of women's work and status in the rural areas of Africa and Asia*, Brighton, Institute of Development Studies, IDS Discussion Paper No 43, p 24.

51. Ester Boserup, *op cit*, chapters 8, 10.

52. 'Women in rural development', Special issue, *Rural Development Network Bulletin*, No 6, Part 11, May 1977, pp 3-4.

53. Ester Boserup, *op cit*, p 221.

54. Adrienne Germain, 1978, *op cit*.

55. *Ibid*, pp 3, 7.

56. Adrienne Germain, 1976, *op cit*, appendix C, pp 11-12.

57. Adrienne Germain, 1978, *op cit*, p 13. The most consistent support for the early programs has been from widows, according to the Women for Women Research and Study Group, *Women for women: Bangladesh 1975*, Dacca, University Press, p 81.

58. Adrienne Germain, 1976, *op cit*, appendix A, p 6.

59. *Ibid*, appendix A, p 6.

60. *Ibid*, pp 7-8, appendix A, p 6.

61. Visit by the author to BEDU small industries workshops, Gaborone and Mochudi, September 1977.

62. Jacqui Starkey and Maryanne Dulansey, *Expanding the external market for Third World crafts — the role of alternative marketing organizations* (paper for Workshop on Third World producers and alternative marketing organizations, September 3-10 1976), The Netherlands, S.O.S. Foundation, mimeo.

63. *Ibid*, p 1.

64. Thanhdam Truong, *op cit*, p 2.

65. *Ibid*, p 8.

66. Donald R Mickelwait, Mary Ann Riegelman and Charles F Sweet, *Women in rural development: a survey of the role of women in Ghana, Lesotho, Kenya, Nigeria, Bolivia, Paraguay and Peru* (published in co-operation with Development Alternatives, Inc, Westview Special Studies in Social, Political, and Economic Development),

Boulder, Colo, Westview Press, 1976, p 96.

67. *Ibid*, p 200.
68. Visit by the author to functional literacy groups of the Botswana Extension College, September 1977.
69. US Agency for International Development, *Project review paper: strengthening women's role in development*, project No 686-0211, Upper Volta, 1975, draft, pp 34-35.
70. Interview by the author with Joy Greenidge, Ouagadougou, October 1977.
71. United Nations, *Assistance in women's activities, Botswana, op cit*, pp 29-30.
72. Donald R Mickelwait *et al, op cit*, p 89.
73. John de Wilde, *op cit*, pp 183-93.
74. Mickelwait *et al, op cit*, p 208.
75. Meeting between the author and home economists in FAO, Rome, October 1977.
76. Interview by the author with people involved in the project, September 1977.
77. Z P Reeves, 'Introducing community development among the Wabena' (Occasional Papers on Community Development 1), Nairobi, 1962, pp 67-68.
78. Interview by the author with Scholastique Kompaoré, Project Director, Ouagadougou, October 1977.
79. Interviews by the author with OCP officials, Ouagadougou, October 1977.
80. Interview with the AID representative, Ouagadougou, October 1977.
81. Adrienne Germain, 1978, *op cit*, p 4.
82. United Nations Development Program, *Activities related to the implementation of national, regional, and world plans of action for the integration of women in development*, paper for the Regional Conference organized by the Economic Commission for Africa, Nouakchott, 27 September—2 October 1977, mimeo, p 2.
83. Interviews by the author.
84. Uma Lele, *op cit*, p 167.
85. UN, *Assistance, op cit*, p 34.
86. Interview with Joy Greenidge.
87. Adrienne Germain, 1976, *op cit*, p 5, appendix A, pp 3, 7.
88. Women for Women, *op cit*, p 81.
89. Marion Ruth Misch and Joseph B Margolin, *Rural women's groups as potential change agents: a study of Columbia, Korea and the Philippines*, Washington DC, George Washington University, Program of Policy Studies in Science and Technology, 1975, pp 91-92.
90. UN, *Assistance, op cit*, pp 37-38. Also visit by the author to the headquarters of the Botswana Council of Women (BCW) and their Gaborone day-care centre, September 1977.
91. Interview by the author with Fatuma Nkhoma and Joyce Kayoka at the Horticultural Training Centre, Kalulushi, Zambia, September 1977.
92. Visit by the author to home economics groups around Kalulushi, September 1977.
93. Visit by the author to Serowe, Botswana, August 1971.
94. Adrienne Germain, 1976, *op cit*, p 5.

95. Visit by the author to the Co-operative Development Program, Ivory Coast, September 1977.
96. Visit to the Co-operative Development Program, Botswana, September 1977.
97. *People*, Vol 5, No 3, 1978.
98. G Warren Povey and George Brown, 'Tunisia's experience in family planning', *Demography*, Vol 5, No 2, pp 620-26.
99. Fatima Mernissi, 'Obstacles to family planning practice in urban Morocco', *Studies in Family Planning*, Vol 6, No 12, 1975, p 422.
100. Swedish International Development Authority, *Women in developing countries — case studies of six countries*, Stockholm, SIDA, 1974, p 51.
101. Peter Fryer, *The birth controllers*, London, Secker and Warburg, 1965, p 254.
102. *Birth control and public health: a report on ten years' work for the provision of birth control clinics*, London, 1932, p 11.
103. Peter Fryer, *op cit*, pp 252, 254.
104. Sam Keeny, 'View from the village', *Populi*, Vol 4, No 1, 1977, pp 7-13. See also Sea Baick Lee, *Village-based family planning in Korea: the case of the mothers' club*, paper for the Conference/workshop on non-formal education and the rural poor, Kellog Center, East Lansing, Michigan, September 26 - October 3 1976, pp 1-4. Also Nepal Women's Organization, cited in *FPIA newsletter*, Vol I, No 4, Summer 1975, pp 1, 4. Also Adrienne Germain, *Women at Mexico: beyond family planning acceptors*, Ford Foundation Reprint from *Family Planning Perspectives*, Vol 7, No 5.
105. Sam Keeny, *op cit*, pp 11-12.
106. *Ibid*, pp 10-11.
107. *Ibid*, p 10.
108. Lester R Brown and Kathleen Newland, 'Abortion liberalization: a world-wide trend', draft for a study by the Worldwatch Institute, p 1.
109. See e g Christopher Tietze and Marjorie Cooper Murstein, *Induced abortion: 1975 factbook* (Reports on Population/Family Planning No 14, 2nd ed), New York, Population Council. Also Christopher Tietze, *Induced abortion: 1977 supplement*. And 'Abortion', special issue of *People*, Vol 5, No 2, 1978.
110. SIDA, *op cit*, p 58.
111. IPPF study cited in *People*, Vol 5, No 3, 1978.
112. See Tietze and Murstein, *op cit*; also Brown and Newland, *op cit*.
113. N Baskara Rao and Ramesh Kanbargi, 'Legal abortions in an Indian State', *Studies in Family Planning*, Vol 8, No 12, pp 311-15.
114. Pierre Gourou, *Les pays tropicaux*, Paris, Presses Universitaires, 1948, p 134.
115. Evelyn Reed, *Woman's evolution: from matriarchal clan to patriarchal family*, New York, Pathfinder Press, 1975, pp 133 and ff.
116. C M N White, *A preliminary survey of Luvale rural economy* (Rhodes-Livingston Paper No 29), Lusaka, Rhodes-Livingston Institute, 1959.
117. See summary in Michael Lipton and Mick Moore, *The methodology of village studies in less developed countries*, Brighton, Institute of Development Studies, IDS Discussion Paper No 10, draft version of

final report, 1972, p 32.

118. Ernestine Friedl, *Women and men: an anthropologist's view*, New York, Holt, Rinehart and Winston, Basic Anthropology Units, 1975, pp 16 and ff.
119. Evelyn Reed, *op cit*, p 139.
120. Mary H Kingsley, *Travels in West Africa, Congo Français, Corisco and Cameroons*, London, Frank Cass & Co, 1965, pp 212-14.
121. *People*, Vol 3, No 2, 1976, p 34.
122. W Parker Mauldin, 'Patterns of fertility decline in developing countries, 1950-75', *Studies in Family Planning*, Vol 9, No 4, pp 75-84.
123. Nancy Birdsall, 'Review essay: women and population studies', *Signs*, Vol 1, No 3, Part 1, Spring 1976, pp 699-712. Ruth B Dixon, *Women's rights and fertility* (Reports on Population/Family Planning No 17), New York, Population Council, 1975. United Nations, *Status of women and family planning: report of the special rapporteur appointed by the Economic and Social Council under resolution 1326 (XLIV)*, New York, Department of Economic and Social Affairs, United Nations, 1975, Document E/CN. 6/575/Rev. 1.
124. Ruth Dixon, *op cit*, pp 5-8. See also W Parker Mauldin and Bernard Berelson, 'Conditions of fertility decline in developing countries, 1965-75', *Studies in Family Planning*, Vol 9, No 5, May 1978, pp 139-41.
125. See Perdita Huston, 'Power and Pregnancy', *New Internationalist*, No 52, June 1977, pp 10-12.
126. Judith Bruce, 'Setting the system to work for women', *Populi*, Vol 4, No 1, pp 10-12.
127. Ruby Rohrlich-Leavitt, Barbara Sykes and Elizabeth Wetherford, 'Aboriginal Woman: male and female anthropological perspectives', in Rayne Reiter (ed), *Toward an anthropology of women*, New York, Monthly Review Press, 1975, p 120.
128. *Population: UNFPA Newsletter*, No 23, January 1977, p 1.
129. Fatima Mernissi, 'Obstacles to family planning practice in urban Morocco', *Studies in Family Planning*, Vol 6, No 12, 1975, p 421.
130. Misch and Margolin, *op cit*, pp 92-93.
131. Richard M Titmuss, *Essays on the 'welfare state'*, London, George Allen and Unwin, Unwin University Books, 1963, pp 94-95.
132. See Jean Medawar and David Pyke (eds), *Family planning*, Harmondsworth, Penguin, 1971, pp 42-43. Also Ruth Adam, *A woman's place, 1910-1975*, London, Chatto and Windus, 1974, p 15. On a similar situation in the United States, see Daniel Scott Smith, 'Family limitation, sexual control, and domestic feminism in Victorian America', in Mary Hartman and Lois W Banner (eds), *Clio's consciousness raised: new perspectives on the history of women*, New York, Harper and Row, Harper Torchbooks, p 123.
133. Titmuss, *op cit*, pp 94-95.

Part Three:
The Effect of Development Planning on Women and their Dependants

Introduction to Part Three

The overwhelming majority of people in almost all Third World countries live in rural areas, practising subsistence agriculture. It is with this sector that Part Three is concerned, and in particular the effect on the women of development planning which, as outlined in Part Two, fails to acknowledge them as participants in the development process. The impact on women of the increasingly dual economy, cash and subsistence, is also outlined as is the contribution of male-biased planning to this division.

The two most vital inputs into subsistence agriculture are land and labour. Chapter 6 examines the question of women's rights to land and their erosion, or attempted erosion, by the planners who allocate land rights to men only. Land is critically important in its own right, but also the key to a whole range of development inputs: credit, agricultural training and extension and so on. Chapter 7 then discusses the question of women's work, or overwork, and the effect of development planning on their workload and their allocation of time between different tasks. Finally, in a concluding chapter the question is raised as to the impact of discrimination against women in development planning on the success or otherwise of projects and programs. Much work needs to be done in this area, but even with the limited data available it would seem self-evident that women respond to economic incentives: that they have a number of strategies for coping with the often intolerable burdens imposed, albeit unwittingly, by the development planners; and that they can be a powerful force for improvement in the success rate of development projects and programs provided they participate fully in the benefits of these.

Chapter Six

Women's control of resources

Control of the land

Of all the resources necessary for subsistence (other than one's own labour), by far the most important is land. In terms of development projects and programs, control of land acquires importance not only in its own right, but as security for credit, and often as providing the criterion for people's access to inputs such as agricultural extension, physical plant, irrigation, chemical inputs and membership of co-operatives. Without land, people lose their security and are reduced to a state of dependence on those with land for the provision of employment. In many countries the colonial experience involved whole populations being reduced to serfdom through their deprivation of land.[1] Similarly, control of land use and its products is of critical importance to women.

Although evidence is far from complete, it seems generally true that in pre-colonial societies women controlled important areas of land; they decided on its use, either alone or in consultation with men; and they controlled the distribution and use of crops grown on the land in question. Ester Boserup suggests that as a general rule, shifting cultivation is characterized by each member of a community which controls a given territory having the right to take land under cultivation. Where women do the farming, it is they who make use of this right.[2] Achola Pala supports this by citing numerous African societies — some of them based on shifting agricultural systems — where women have important rights to land and to the crops grown on it. In many cases, there would be a distinction between women's fields and men's fields, and between women's and men's crops.[3] She also stresses the rights to land and its produce among women of transhumant groups, as well as pastoral women's ownership of cattle and other livestock, their milk and other products.[4] Rights were often qualified, for women as well as men, by the rights of others to the same thing; thus women might have cattle consumption rights, but not the right to sell them.[5] Outright ownership in the Western sense would be an inappropriate concept to apply here.

The distinction between the ownership and control of land and other resources is a vital one. Excessive interest in the right to transfer land to others, which is of relevance mainly in the context of establishing theoretical descent lines, could be misleading in terms of the actual decisions about the land and its use, which are of immediate importance in the farming system. It also tends to assume that land is controlled by one person only, whereas this may be the exception rather than the rule. For example, on smallholdings in Jamaica one of the main characteristics of family land is that it is left jointly to two or more members of a family. When a married man dies, his widow takes over control for as long as she lives. Only blood descendants of both partners normally inherit rights to land, except that a man's illegitimate children and other 'outsiders' may be included if they live in close association with the family and its land. A child living away and neglecting the parents may also forefeit rights to the land.[6]

The fact of land being controlled by more than one person, in some kind of partnership, is one that must be kept in mind when dealing with the idea that 'traditionally' only men own land. Since, in many different societies, they are in fact given land only at marriage, there is a clear implication that the woman's participation is essential to its proper management. The same may be true of women's rights to land, acquired on marriage, as in a 'binna' (matrilocal) marriage in Sri Lanka.[7] The emphasis is on a viable partnership in management of land and its eventual transmission to the next generation. Rockwell suggests that the co-operative production system with its division of labour ensures and reinforces the prolonged co-operation of the marriage partners, which is often reproduced in ritual and ideology: the fertility of the human couple is compared with the fertility of the fields under their joint management, or to that of their livestock.[8]

Similarly, marriage is generally characterized by an exchange of commodities, between the families as well as between the two people concerned. Among most cattle-owning peoples of Africa, for example, marriage is legalized by the transfer of cattle from the husband's lineage to the wife's.[9] Many of these exchanges are conditional, however; if the marriage breaks down, for example, through failure to produce children, disputes between the partners, or failure of either to contribute satisfactorily to the household's subsistence, then some or all of the exchanges may be returned. The marriage partners themselves may forfeit control over important resources; for example, Audrey Richards points out that among the matrilocal Bemba, the use of millet from a field cleared for a woman by her husband, as part of the marriage contract, was hers only as long as the marriage lasted. Upon divorce, it was often the husband's female relatives who commandeered the grain.[10]

In many cases where the couple's land rights revert to the husband, a woman may have the option of claiming another piece of land in her home village, particularly if she has been careful to retain that option by frequent visits there, and by gifts to her own family. Vanessa Maher suggests that this right is an important factor where there is a high level of interaction among female kin, and a general instability of marriage. Women's rights vis-à-vis their brothers and fathers may well be more advantageous than in the case of their husbands, and they will then demand high standards in their marriage relationships.[11] In this way, women's common interests and support for each other give individual women a much greater degree of choice in marriage.

Women may also use reserves, in the form of gold and jewellery, as security in the face of what is seen as the uncertain subsistence offered by marriage. In Tunisia, for example, jewels are passed down from mother to daughter and provide an important source of cash and some independence of their husbands' control.[12] Among Malay peasants, where marriage is also seen as a fragile arrangement, women control the family gold, its main reserve in case of a crisis. They decide what it should be used for, and with support from other women would strongly resist pressures to use it for a feast, or to finance purchases or economic ventures by their husbands. Women are concerned to provide for the economic security of themselves and their children by acquiring gold and land registered in their own or their children's names, and expect husbands to provide these resources as an assurance that they are satisfied with the marriage, including the women's work in the rice fields and the home.[13] Women may also, in many cases, control land which is formally owned by men. In Ghana, for example, many men leave their cocoa farms in the full charge of their wives and, according to Polly Hill, 'as the effective (if not the legal) custodians either of their husbands' or of lineage property they are apt to exert much power . . .'[14] In Morocco, as in much of the Middle East, women quite often take over management of the household farm when their husbands die, depending on a female network in order to do so. Many tasks taken over by them, with the help of other women, are usually men's, although they are not all done in the same way.[15] This kind of situation is obviously very common in areas of heavy out-migration by men, where women take over management of land which is in the men's name, often supporting each other in order to do so.

The Western concept of outright ownership by a single individual as the basis for land tenure has had the effect of suppressing the practice of joint ownership, or qualified rights to land and other resources, in which women participate. Since colonial and development planners have been extremely reluctant to recognize women as holding rights in land, except perhaps in matrilineal or matrilocal descent systems, the right to land and control over its use were almost invariably ascribed to men.

The changing definitions as to rights over land and other resources are of crucial importance to women, and to overall development. Achola Pala suggests that in pre-colonial pastoral or agricultural economies, women were usually well protected economically because their 'usufructuary' rights in land and cattle were well defined, and were actually more effective than individual ownership. The normative emphasis on usufrust extended also to other resources such as fish, game, salt licks, water, herbs, vegetables, fruits, fuel, clay and thatch. This favoured the individual economic rights of all kinds of people. In addition, since productive labour held precedence over formal ownership, the system guaranteed control over the products of the land and other resources, to all those who were working.[16] John de Wilde also stresses the importance of the traditional rights to land in a developing rural economy: land, he states, 'is at the heart of the African's security and many of his [sic] traditional practices and customs'.[17] He argues in fact that usufruct is not a separate category of rights but closely related to ownership, although control may not be absolute as in cases where the right-holder may not transfer the land to outsiders. In most of tropical Africa, individual families may be said to control their own land, together with the right to transmit it to successive generations, at least as long as they use it. Distinctions between individual and group rights are of no practical importance unless land is becoming scarcer or more valuable.[18] Ester Boserup suggests that the concept of 'ownership' is ambiguous in relation to tribal tenure and is often used by chiefs, who have formal although perhaps severely qualified control over land, in order to justify a claim for ownership rights in the whole tribe's land, when this acquires a cash value. Similarly, she argues, men can use the European con-

cept of ownership to give the impression of an absolute right
over land which may not in fact exist in the community in question.[19] This obviously would be readily accepted within the
ideology of a Western culture that expects men to control their
own activities, and women to be controlled by men. As George
Dalton points out, a Western description of economic transac-
tions in a 'primitive' society, instead of describing a Trobriand
Islander man giving yams to his sister's husband partly in recognition of her rights to land he is using, would say that the man
is 'maximizing prestige'.[20] Similarly, women's rights to dispose
of crops by home consumption, retention as an emergency
reserve, exchange, and gift-giving are all unlikely to be recognized in economic terms; particularly as they do not involve
cash and do not fit into the categories of economic description
favoured by development planners, as outlined in Chapter 4.

Land transfers from women to men

In many cases, then, women lose their rights to land because
these are not recognized within a simplistic male ideology of
outright ownership. However, if this principle were applied
uniformly and without further distinction between women
and men, logically there would be some areas where women
acquired such ownership. Most obviously, this would apply to
systems of matrilineal descent and inheritance.

Matriliny has proved a disturbing issue for Westerners, for
whom, in the words of Rattray, it brings 'results which appear
to us either unnatural or grotesque.'[21] There is an overt hostility
to matrilineal systems which is based apparently on the idea
that men must control women. This seems to be the argument
of Claude Lévi-Strauss, a leading structural anthropologist who
insists that men are uniformly in control of all societies, and
that there is therefore an 'absolute priority of [patrilineal]
institutions over matrilineal institutions'.[22] This idea has frequently been repeated by others, including Maurice Godelier,
who seem content to accept the idea as an important principle
of social organization despite the lack of evidence other than
the supposed 'natural' and universal control of women by
men.[23]

This position is by no means confined to academic analysis;
it is reflected in conscious policies to eradicate matriliny altogether, in the name of development. Mary Douglas cites a number of Western men who claim that matrilineal descent is com-

patible only with poverty, and is inappropriate for economic advance. It is also seen as an unnatural system full of intrinsic conflict, because of the alleged disruption of the 'elementary family' — apparently meaning the Western-style nuclear family of male 'head', his wife and their children.[24] Mary Douglas argues that, far from matriliny being a conflict-ridden system and a mirror-image of patriliny without its advantages, it in fact offers a number of benefits. These are of particular relevance to situations where egalitarian distribution is essential to survival, as a result of successive gluts and shortages in the subsistence base. As a system which cannot sustain long and clear-cut descent lines, it gives scope to the enterprising individual to override ascribed roles rather than being limited to a set hierarchy; it relies on the introduction and assimilation of new members, giving it a remarkably open quality; and it is well adapted to the distribution of food in a system of general transfers. Furthermore, she argues that matriliny, far from being incompatible with economic advance can, like any extended family system, offer numerous advantages as a base for rapid economic expansion: matrilineal clans have introduced new systems of production with members participating rather like shareholders in a new venture. She suggests that the spontaneous decline of matriliny is related more to economic recession and shortages of resources than to expansion.[25] It would also seem reasonable to suppose that extreme inequality of distribution of the benefits of economic expansion may be a feature of patrilineal much more than matrilineal or bilateral systems. Although Douglas seems to accept the assumption that economic development must involve economic differentiation, a serious check on this tendency might be one of the most valuable contributions to be made by matriliny, and one which is compatible with the more recently advocated ideal of development as promoting the welfare of the whole population rather than the favoured few.

This connection appears not to have been recognized by planners. In fact, since early colonial days Westerners have been opposed to what seems to them the 'unnatural or grotesque' system of matriliny. A colonial power can effectively change a bilateral or matrilineal system of inheritance virtually overnight by instituting a new, Western-style code and land tenure system; this process has been described in the case of Jamaica by Edith Clarke, for example.[26] Evelyn Reed cites examples of Westerners witnessing the dramatic reversal of kinship systems 'before their eyes'.[27] The change extended also to residence, and she sees Christian missionary influence as having been strongly in favour of virilocal marriage.[28]

The struggle continues. As Polly Hill remarks, 'outside observers of the matrilineal scene have never ceased forecasting its imminent collapse.'[29] There are powerful interests invested in this collapse: an unspoken alliance between the Western-orientated planners and the male élite. The first men to opt out of the matrilineal system in Ghana, as Mary Douglas points out, were professional men, ministers, careerists, clerks and Christian farmers.[30] Among cash-cropping farmers in Mwinilunga described rather revealingly by Turner:

> 'Farm heads [sic] were disencumbering themselves of many of the obligations of kinship, and retaining for their own use and for the use of their elementary families money they earned as wages and by the sale of cash-crops or surplus subsistence crops.'[31]

In other words, once defined along the Western model as 'head' of the nuclear family, and in full control of its land and the crops produced from it, a man could defy the distribution system of the matrilineal extended family and begin accumulating wealth for himself and those defined as being under his jurisdiction.

The strength of the anti-matrilineal forces is obviously considerable; however, to pronounce the system extinct, as many planners do, may well be premature. There is great eagerness to jump to conclusions about this on dubious evidence: for example, in Ivory Coast where a survey had been conducted as to people's view of inheritance, I was told by the survey leader of the institution concerned, the Institut Ivoirien d'Opinion Publique, that the result showed the breakdown of the matrilineal system in the areas concerned.[32] The result, however, is ambiguous, since a variety of different areas were covered, and the question was, 'According to the present law, who should you inherit from in principle?' Since the Ivoirien legislation clearly specifies, along the lines of France's Napoleonic Code from which it is derived, that inheritance should be from the father and not the uncle or anyone else, the 54 per cent who named the father were as likely to be showing knowledge of the law rather than personal preferences, especially in view of the familiar tendency of respondents to give answers that they think will please survey-takers. Thirty-six per cent said that inheritance should be from the father and mother; however, this is also ambiguous, since inheritance from the mother is perfectly compatible with matrilineality. Some respondents could have had in mind the mother and her family, including her brother.[33] In a related study of an area heavily involved in the development of sugar, the survey-takers' outright hostility to what they found is clear. Throughout this study the strength of matrilineal ties is apparent; for example, women give money earned at the sugar factory to their mothers and uncles, rather than their husbands and fathers.[34] This is seen as contrary to the desired objective:

> 'The attachment of a married woman to her family of origin and matrilocal residence constitute brakes on the economic partnership of the married couple. Hence the anxiety of those [men] who have not found employment at Sodésucre and their feeling of having lost [sic] their hold over their wives.'[35]

Presumably these men never had any hold over their wives in the Western sense, and so could hardly have lost it; however, encouraged by powerful Western influence in which male control of wives is the norm, they may well have come to feel strongly resentful at the contrary situation. Certainly the development researchers and planners seem to be doing their best to support the idea that a husband should have 'a hold' over his wife.

The World Bank and the suppression of matriliny

Given the resilience of the matrilineal system in certain areas where it has been proclaimed extinct, the current policies and practices of development projects and programs operating in these areas may be very important in determining the final outcome of the conflict of interests. A clear example of this is the World Bank's model project in Malawi already mentioned in Chapter 4, the Lilongwe Land Development Program (LLDP).

The local management of this project can hardly claim that there was no information available about matriliny as a factor of land tenure in Malawi. Under British colonial rule in what was then Nyasaland, Village Re-organization Schemes were devised as an integrated program for land consolidation and technical improvements in land use, which by 1958/59 had been extended to 38 different areas covering some 200,000 acres. During the next few years, however, all the schemes were abandoned due to opposition from the local people, and the situation reverted completely to the *status quo ante*. Among other miscalculations it was apparently assumed by the colonial Department of Agriculture that the consent of the chief of the area was enough, and that he could and would enlist the genuine support and participation of the headmen and the people. John de Wilde comments that there seem to have been particular problems in winning over the women who, in the matrilineal society prevailing in the affected areas, were of critical importance in any land reform and who were sometimes 'needlessly alienated' by the way in which land redistribution was carried out.[36]

Management policy at LLDP is of particular significance because it lacks the excuse of ignorance about the hazards of trying to suppress matriliny, and specifically women's rights to land. The Program also boasts an evaluation section which conducts surveys designed to provide information on the context of important policy decisions. The LLDP is also of major consequence to rural development policy throughout Malawi, and for World Bank rural development ventures in general. In Malawi a succession of big rural development projects, financed by Britain, Canada, the European Economic Community and other major donors, have been set up more or less using LLDP as a model.[37] LLDP, according to the World Bank, is significant in being aimed at the poorest farmers, the 24 per cent of the population cultivating less than three acres and earning half the national average.[38] As such, it is a model for the Bank's 'new' approach to rural development, stressing benefits to the smaller farmers and an 'integrated' approach. Another Bank document states: 'The significance of LLDP for the IBRD is that the project typifies one of the Bank's two principal strategies for what are called "integrated smallholder development programs". . .' Instead of focusing on a single cash crop, it aims to include a wide range of activities that support production both for subsistence and for the market.[39]

In many ways, of course, within the LLDP there is no more understanding of the workings of a subsistence-based economy than there is in the average project. People in the program area are to be guided towards 'a cash orientated way of life' involving increased sales through official marketing outlets — government marketing boards which offer extremely poor returns to smallholder producers. There is thinly disguised contempt for the co-operative systems of production and distribution prevailing in the area. 'Rural peoples' are considered intrinsically conservative and resistant to change.[40] One evaluation document states:

> 'The major obstacle to change is the socialistic [sic] principles underlying traditional tribal life. No man should rise too high above his fellows in terms of material wealth, and the natural counter-

weight of the extended family, with its limitless responsibilities and obligations ensured that this did not happen . . .

Survey questions attempting to categorize people as 'traditional' or 'modern' equate social responsibility and altruism as 'traditionalist' and implicitly undesirable.[41] Or as one senior UNDP representative put it, 'Il faut déraciner les idées, les coutumes. . .'[42]

LLDP surveys[43]

Perhaps foremost among these supposedly undesirable customs is matriliny and the widespread system of 'socialistic' distribution among the extended family which appears to go with it. The management has apparently determined, on rather flimsy evidence, that matriliny is as good as extinct. In an early document which deals specifically with the question of matriliny, they cite survey responses from the people they define as 'growers' (mostly men):

> 'The survival of Matriliny in Africa is a talking point among Sociologists and there is certainly considerable evidence of an increased desire amongst growers that inheritance should pass to their children. Growers were asked who would cultivate their gardens after them and 73 per cent stated that this role would pass to their own children.'

The problems with this rationale are the same as with the Ivory Coast survey described above: the answers are likely to have been influenced by what the enumerators (themselves very Western-orientated, with salaries paid by the project) wanted to hear. In addition, the question of what was understood by 'their own children' is an open one: if, for example, a man felt his nephews or nieces to be his own children, and entitled to inherit his goods, he would probably have responded in the affirmative. There is also the element of uncertainty about what those women defined as 'growers' may have meant: their property would pass to their own offspring regardless of whether succession followed matrilineal principles. There is no breakdown by sex of the answers to this survey. (Elsewhere, however, in a document of two months later, it is stated that 97 per cent of male respondents and 91 per cent of females said that their children would inherit; this appears completely inconsistent with the previous composite figure.)

There is further uncertainty in that much stress is placed on the fact that few respondents specified inheritance as passing from the maternal uncle to his sister's children. This is consistent with a view of matriliny espoused by Lévi-Strauss and others, which insists that men, as uncles rather than fathers, are still in control of the system. However, where women in fact hold land in their own right, and if their husbands' land rights are contingent on a continuing marriage and matrilocal residence, it would be of little interest what children stand to receive from their maternal uncle (who could well be living in a different area, with his wife). The childrens' most important link with land is through their mother and her family, who control the land on which they work. The daughters remain there, and are given the use of land; the sons work on their wives' land, and their own rights are very strictly qualified by, among other things, a fairly lengthy trial period during which they have to demonstrate a capacity for hard work.

The system is illustrated in considerable detail in one of LLDP's own evaluation documents. It describes the system somewhat emotively as 'matriarchy', and stresses the almost universal practice within the old tradition of a young man going to live in his wife's village; during that time the man is known as *mkamwini*, and his period of residence *chikamwini*. The husband's position in the village and his access to land was determined by his wife's kin. On completion of the probationary period, which used to last up to two years, he would be given land to farm on his own, and as long as the marriage lasted his rights to this land were inviolable; if it did not last he had to return to his own village and seek land there. The practice of *chikamwini* was described as 'still widespread amongst the people of the area', but no longer the rule. The practice was said to be changing gradually from a lifelong commitment into a formality lasting only from a few months to two years — although it is also noted that, of the male growers surveyed who were currently on *chikamwini* (26 per cent of all male growers), 63 per cent had already been there for nine years or more. Where a husband and wife later moved to the husband's village, they were likely to be given land by his kinfolk. It would appear that the majority of male growers living in their own village had done at least six months of *chikamwini* — 67 per cent according to the social survey. The picture is confused by the fact that for many couples, wife and husband both belonged to the same village; a period of absence for *chikamwini* may therefore have been unnecessary even in terms of traditional obligations. The document itself states that the 31 per cent of married male growers who had done no *chikamwini* at all is probably an overestimate for this reason. Elsewhere, survey data of about the same date showed that only 21 per cent of the gardens belonged to growers who were practising uxorilocal marriage (where a woman moves to her husband's area).

This data on where people actually live and farm seems to be completely inconsistent with the claim that matriliny is virtually extinct. One may venture to suggest that simple factual questions regarding residence are likely to be much more reliable than largely hypothetical ones such as who would inherit property, assuming that ownership had any real meaning. The evaluation document, already cited, implicitly recognizes the inappropriateness of the Western concepts of ownership and inheritance when it outlines the philosophy behind Chewa land tenure:

> 'Under Chewa custom land is not a marketable commodity. An individual may use a piece of land for many years, but he can never "own" it. Possession of that land transcends his lifetime; it belongs to the dead, the living and the unborn. The individual, therefore, does not own a piece of land, but has "rights" over it. Authority to control and arbitrate on rights to land is traditionally vested in the Village Headman . . .'

This helps to put in context the apparent contradiction between responses to survey questions which are interpreted as showing the breakdown of matrilineal tradition, and residence patterns fully consistent with its continuing vitality. In practice, cultivation rights (as opposed to ownership) are passed on from parents to children, although it is often, as the LLDP document points out, a 'moot point' as to which parent is actually handing on those rights. Since they hold joint rights to the land, it may be a distortion to see either individual as passing on the rights to it. Some growers with a father who is *mkamwini* may name him as the holder of the rights, if asked by an enumerator to identify a single individual. The whole survey was also greatly distorted, as the document itself admits, by failing to consider other relatives still alive at the time of 'inheritance'; so some respondents might name the father as passing on the cultivation rights because the mother had already died. Under the traditional system, cultivation rights could also pass from parents to children well before the parents died. Twenty-three per cent of all growers in fact named the village headman, the formal arbiter of land questions, as the source of their land rights, a higher proportion than those (21 per cent) naming the father. Other answers covered a wide range, including maternal relatives, (10 per cent as against two per cent for paternal relatives), spouse's relatives, mother's brother (seven per cent), mother (17 per cent) and 'other' (10 per cent). Several of these are indicative of matrilineal tradition, as is the fact that many more men than women had land rights derived from 'spouse's relatives' (13 per cent and four per cent respectively).

Land Registration at LLDP

Despite evidence contained in their own evaluation documents, LLDP management is carrying out a program of land 'registration' which conflicts not only with the matrilineal tradition, but with the whole concept of joint interests and widespread distribution networks that go with it. The program is based on Malawi Government legislation of 1967 which provides the legal and institutional framework for the classification and registration of land — a law that is of primary benefit to the handful of men who are emerging as large landowners and who are often also members of President Banda's Government. The official objectives of the LLDP program make it clear that it too will act to entrench individual 'security' in absolute ownership of land (regardless of use), which in turn will guarantee access to a variety of development inputs offered by the scheme. All except the clan or family 'head' registered as landowners will be expected to become dependant, losing all their own rights to the land. The land allocation program:

> '. . . provides for land security, which did not exist for many under the traditional system within the matrilocal, matrilineal society of the Achewa people; with greater security, sound investment can be made in property . . . a system of land holding is instituted with the legal protection and definition of boundaries . . . ownership of land enables it to be used as a security against grants and loans; and transactions can be made according to a required system; land can be consolidated into family unit holdings . . .'

In an area where the vast majority of people have been, or are involved in the matrilineal system of *chikamwini*, this is tantamount to an all-out attack on a people's way of life.

It would, of course, be feasible to change from the traditional land use system to a Western 'ownership' pattern without depriving women of their security as farmers. Indeed, a system genuinely promoting the idea of a 'family unit holding' should logically grant ownership rights to the family, involving all the adults working on the holding as well as provision for the children. This, however, is categorically excluded as a feasible option in the LLDP approach: only one person in each 'family unit' will have legal title to the land. This person is variously described as the 'family leader' or 'family representative' and similar euphemisms, with the claim that all members of the family participate in the decision as to who this should be:

> 'This family unit land is categorized as "private freehold land" with the elected family representative becoming the absolute proprietor of the land.'

After the appropriate determination:

> '. . . a land certificate is issued to the proprietor(s) of the private freehold land, the family leader being the elected representative of his [sic] people.'

No procedure is laid down for a family 'election', nor is there consideration for the fact that election normally involves the right to remove a representative who acts against the interests or wishes of the people. Thus the prospect of an unprecedented family democracy is undermined by several factors making it likely that the 'elected family head' will be a man, following the familiar 'head of household' model. The definition of family, however, is a strange one, covering much larger numbers of people than even an extended family. In two of the areas, holdings have been consolidated into between 150 and 250 acres, with no internal boundaries; this contrasts with the average holding of a peasant family, which is only 4.5 acres. The new pattern is very much like the plantations already set up by the favoured few in Malawi, which have robbed tens of thousands of people of their land.

In LLDP surveys involving growers' claims to positions of traditional authority, the post *'mwini mbumba'*, the head of a clan unit of which there are usually between two and four in a village, was claimed by six per cent of the female growers and 26 per cent of the men. This appears to be a gross overestimate on the men's part, since one would normally not expect to find more than 15 per cent of them acting as *mwini mbumba*. The question may have been misinterpreted as referring to *mbumba*, which for most adult men means their responsibilities under the matrilineal system to their sisters and their sisters' children. This minor misunderstanding may indicate the kind of problems that would arise in defining the 'family representative', and it would be ironical if men used their responsibilities to their sisters, under the matrilineal system, to claim control over their wives under a system explicitly designed to eliminate matriliny.

The registration program is also likely to rely on existing project definitions as to who is or is not a 'grower', which as shown in Chapter 4 heavily favours males, even if they are not actually present or growing anything at the time. Men are deemed by the project to be 'growers' with rights to land held by a relative, wife or wife's relative. The latter two categories account for 37 per cent of the total for male growers. The grower registers, based on such arbitrary assumptions, provide the basic data for policy implementation. If land registration were to follow the principles of the grower registers, this would mean a substantial transfer of land from women to men — many of the men being absentee landlords with their wives working virtually as unpaid labourers. This would affect particularly the large number of women in polygamous marriages. About 20 per cent of men are polygamous; percentages for women are not provided, but since the majority of these men have two wives, approximately double the number of women would be involved. Except for village headmen, the wives of polygamists, particularly the second and subsequent ones, most commonly stay in their own villages. Even where a woman has her husband living with her, he may be considered to have no authority whatever over her garden, being *mkamwini* (i e working a probationary period on the woman's land); these cases account for 13 per cent of all registered female 'growers' (and there may be others where enumerators arbitrarily list the husband as holding decision-making authority). Many gardens where this is the case were probably excluded from grower surveys altogether: 'The surveying of land was restricted to gardens in which the sample grower was the managerial decision maker . . .'

It would be surprising if land registration did not correspond to a considerable degree with the categories set up for grower registers, although efforts are being made to introduce at least an attempt at popular participation in the decisions. Introductory meetings set up by project officials elect a 'land committee' to advise the project's demarcation officer and represent the interest and claims of all parties. Notices are then distributed throughout the intended area of work giving notice of commencement of the program 'so as to allow all interested parties and those with claims to attend.' No mention is made of exactly how meetings are gathered, although it must be relevant that in this area, as in others, many more women than men are illiterate, and so less likely to participate in the procedures which are advertised by written notices. Only 17 per cent of male growers' wives and 22 per cent of the female growers have any formal education at all, compared with 55 per cent of male growers. Only 14 per cent of female growers are at all literate, compared with 53 per cent of the males.

The contentious question of the boundaries between the 'various family groupings' is settled with the approval of the village headman — usually a man with several wives in the same village, and therefore with a strong vested interest in male-centred registration. Only a few village 'headmen' are women (one per cent of the survey total). Also participating in the decision on the boundaries of a family grouping are the so-called 'family leaders' (although it is not explained how a 'leader' can be elected before the determination about whose land is to be included is decided). In any boundary disputes, the land committee is the final arbitrator. Since non-traditional authority, particularly the Malawi Congress Party (MCP) is of considerable importance, these official structures may be expected to have substantial influence on the composition and decisions of the land committee. Women have very little representation on Village Planning Committees or Local Education Committees in the surveyed areas, and only two per cent of the women belong to the MCP as compared with 12 per cent of the men.

The officials responsible for the land registration program are probably acting with the best of intentions, sincerely believing that, as their economics text books would tell them, people with secure land title have greater incentive to invest in the land and improve its productivity. The problem is that, in drastically reducing the complex network of land rights held by a variety of people to one certificate of ownership per 'family unit' and effectively dispossessing all the rest, particularly the women, many more people lose such incentive as they had, than gain something they really need. All the family members are, in theory, supposed to be named on the register, but many women, already excluded from the growers' register, are likely to be omitted. Even if they are listed, it is hard to see how they could enforce their interests. The problem is compounded by the fact that, for the new title-holders, the incentive to invest their own labour in their land is actually reduced; whereas previously a man was obliged to work on a plot in order to retain rights to it and to living with his wife, this is now abolished as a condition of ownership. Given the very high incidence of absentee husbands because of polygamy, migration to South Africa and Rhodesia, and other reasons, many plots are likely to receive little or no attention from the official owner, while all those working on them will have no incentives derived from ownership.

Even in terms of the official objective of increasing investment, therefore, the land registration scheme would seem
counter-productive. In terms of the people's interest in sustain-
ing a form of social and economic organization that provides
maximum security in a subsistence economy, the scheme could
be a disaster. It is not surprising that, as one document mentions in passing, there is evidence of 'suspicions and emotions
engendered in dealing with land changes.' Unofficial observers
speak of a virtual state of war in the areas concerned.[44]

Taking women's land: a world-wide phenomenon

Africa

The process of removing women's rights to land is by no means
confined to Africa; however, it is there that it has been observed
most frequently. This appears to be because land tenure in Africa
has been, until very recently, the least affected by serious over-
crowding (apart from the 'reserves' set up in countries where
numerous white settlers were taking over large tracts of land,
as in parts of Southern and Eastern Africa). The impact of
Westernization is also much more recent than in much of Latin
America and parts of Asia, and has been recorded by observers
since the Second World War.

Ester Boserup, for example, notes a number of cases where
European colonial administrations instituted land 'reforms'
which eliminated women's rights, most conspicuously in areas
of predominantly female farming as in the French and Belgian
Congo, the Bikita Reserve in Rhodesia — where a 1975 land
reform excluded married women from all rights to land, while
often their absentee husbands received formal title — and in the
Transkei reserve of South Africa.[45] John de Wilde describes a
land consolidation and registration program conducted by the
colonial government in Kenya, 'avoiding needless refinements'
as an economy measure and relying heavily on 'committees of
landholders representing, above all, the men [sic] who under
customary law were competent in land questions'. The authorities provided 'surveyors' who were the ultimate arbiters in
disputes[46] — only males then being eligible for government
employment. De Wilde, however, questions the advisability of
consolidating land holdings.[47] With the loss of women's customary rights and with no obligation on right-holders to work
their plots, he also notes that in the Central Province, where
registration had been carried out against a background of local
enthusiasm, 'we noted considerable areas which have been left

idle or virtually idle', to the extent that special fiscal penalties were having to be devised to penalize men who held land without working it.[48]

Colonial land policies in Tanganjika are described by Reynolds as following a similar pattern to those in Kenya; the colonial authorities, equating the power held by certain traditional male leaders to allocate land with the Western concept of ownership, assumed that men owned all the land; in areas of land registration, therefore, 'the power to allocate was interpreted as the registrable interest'.[49] As men were given formal title to land, women lost the guarantees of the traditional system, and could not, in many cases, prevent it being sold to complete outsiders. Furthermore, she points out, there was no functional institution to support the new system of individual ownership rights, to provide for competing ownership claims equivalent to those offered women by the original system.[50] Erich Jacoby, although he completely ignores the role of women, points out the factor behind the colonial policy of individualizing ownership rights which is of central importance here: it encourages manipulation and speculation in land by foreign commercial interests and also by those who can use a traditional position to claim private ownership, such as chiefs and headmen, who may exploit the situation to become feudal lords or capitalist farmers. He stresses that the 'typical colonial trend' of individualizing land tenure continues under the present governments, 'trained in the spirit of the colonial powers'.[51]

Asia

In the case of Asia there is much less evidence available about drastic changes in land tenure systems. In India at least, colonial rule seems to have meant the sudden freezing of a rapidly evolving system of land tenure at a time of swiftly rising land prices; it involved an effective alliance with the big landowners, according to Margaret Haswell.[52] Guy Hunter describes the currently prevailing land tenure systems of *zamindari* and *ryotwari* as those imposed by the British colonial administration: the former adapted from the preceding Moghul system but involving some misunderstanding of that system.[53] Thus, Indian land tenure is a strongly colonial institution, subject to the biases, including male bias, current in Western ideology from the start of the lengthy period of British colonial rule there. It is ironic that in moves to evade efforts at land reform by the Indian Government, large landowners have registered land in the name of the women in their families. The difference between this transparent device and any real control by women over the land in question is evident from the fact that registration has also been practised in the name of unborn children, household pets, and farm servants who even remained unaware of their names being used in this way.[54]

Land tenure in other Asian countries is probably still much more flexible than in India, with many of the women being recognized as having rights to land. In Ceylon after the Second World War, for example, a colonial settlement scheme selected only men as registered participants (although they had to be married, and have at least two children over eight). When required to nominate a successor, however, there were as many women as men named. Out of the 43 colonists 19 nominated a son, 13 a daughter and seven a wife.[55] In South East Asia, colonial and post-colonial administrations have apparently transferred women's land rights to men in a similar way to that observed in Africa. Among matrilineal clans in colonial Malaya, for example, much land was traditionally passed down from mother to daughter. Under the British administration, however, the traditional code was embodied in the new law, but only with regard to land that was recognized as falling under clan control when the land titles were registered; this, in practice, was almost entirely rice-land, together with the plots used as the housing site. The subsequent development of cash crops from rubber and fruit orchards is outside the bounds of the now inflexible traditional code, and women are therefore largely excluded. Only the matrilocal marriage pattern and other characteristics of the still vigorous family system, together with the fact that rice-land and house sites are not particularly sought by the men — who expect to use their wives' land — keeps this kind of land in women's hands; in the event of their acquiring significant value there would be no legal sanctions available to the women for enforcing their rights.[56]

Latin America

In Latin America, women have even less claim to land based on traditional systems of land tenure since, as has been pointed out by André Gunder Frank, land was the focus of increasing demands by settlers and commercial interests as the value of its products increased. It was acquired with little or no respect for any body of law by *merced* (grant), conquest, expulsion of Indians from their communal lands, and, later, also of mestizo and even white homesteaders from their private lands:

> '. . . claims being staked out and later legalized through bribery and/
> or falsification of documents, often through purchase or in default
> of debt payments by the previous owner, and through a variety of
> outright fraudulent means . . .'

He stresses the fact that the system of land tenure was based on the exchange of property titles, favouring those who had enriched themselves in commerce, especially mining, and tending to a slow process of land accumulation in the hands of a few. Frank sees the 'cash nexus', rather than any aristocratic or feudal traditions, as having ruled in Latin America from the very beginning of colonial rule there.[57]

Perhaps the only comprehensive survey of trends regarding women's rights to land in Latin America is the study by Patricia Garrett of the situation in Chile from 1935 to the present. Under Chilean law, a woman's property passes to the control of her husband on marriage, unless expensive legal precautions are taken which are clearly not feasible for peasant families. The law of marriage is closely modelled on European concepts in this respect; a relevant description of marital rights in the 1964 Code indicates that 'the *potestad marital* is the set of rights that the laws grant to the husband over the person and goods of the woman.'[58] Faced with a virtual ban on ownership of land, combined with a rapid loss of jobs in agricultural wage-labour on large estates, women 'were increasingly confined to the smaller farms as unremunerated family members' totally dependant for any income on their husbands. It is 'practically impossible' for a woman to maintain herself and a family in the Chilean countryside without economic support from an adult male, whether husband or son.[59]

The dilemma of women in situations that allow men to monopolize land dealings is forcibly expressed by Lily Poznanski of the Solomon Islands, in the Pacific. Explaining that the current law is a combination of customary, unwritten law and formal legislation modelled on the British system, she adds that in islands like her own, Ysabel, women have had a strong position in society with regard to land dealings and other issues:

> 'But the modern and foreign influences are somehow shadowing our
> women in some good customs, for example those on land dealing.
> This is due to lack of education, the dominance of men in our deal-
> ings with foreigners, and the lack of presence of our women in the
> legislature. So, although by custom women should have these powers,

we seem to be losing them, although our menfolk keep telling us that they are trying to strengthen custom.'[60]

Cash crops: agricultural development for men

Colonial and post-colonial law and administrative policy have been powerful forces for the erosion of women's rights to land. Other powerful influences have also been involved, in particular the concentration of agricultural 'development' innovations on men and the increasing identification of women with a stagnant and dwindling subsistence sector in an emerging dual economy.

The involvement of men in official efforts to develop cash cropping, especially for export, has been observed by a number of writers particularly in the case of colonial Africa where, as William Allan sums it up: 'The extraordinary development of the production of cash crops by African cultivators, mainly within the last 40 years, is something unprecedented in colonial history'.[61] It would be misleading to suggest that only men grow cash crops; in fact women both grow and trade cash crops to a considerable extent, as Sherilynn Young stresses, for example, in the case of Mozambique.[62] Ester Boserup notes one case where administrative efforts to promote cotton production in Uganda brought an enthusiastic response from many of the women; deliberate efforts were then made to turn the crop over to men.[63] In Gambia, where there was great concern in the late 1950s over low levels of rice production, the important role of women in this area was noted, and defined as an obstacle to making this a cash crop, in an official report for 1958/59:

> '. . . in order to give a further fillip to rice production and to raise its status as a woman's crop . . . Lantern slides were shown in the villages, and in 1957 young men were accepted as trainees at the Jenor station to acquire practical knowledge in the correct method of rice cultivation.'[64]

Despite the large subsidies and an extensive training program, the effort to replace women with men in rice production was largely ineffective in that case.[65]

A variety of different administrative measures have been applied to promote cash-crop production for export as a male-controlled enterprise in smallholder agriculture. Reynolds includes coercive extension services, marketing controls, forced cultivation, taxation aimed at men, and the provision to them of more sophisticated inputs and equipment. A new dual economy, she suggests, was created deliberately.[66] A review of the situation by SIDA (the Swedish International Development

Authority) concludes that any planned innovations involving new techniques are always introduced to men, and that this demotes women from independent producers to labourers with no rights to a share in the profits of the enterprise.[67] In certain situations, such as the formation of *ujamaa* villages in Tanzania, innovations which are of little practical value in increasing production, such as tractors, would be offered as an inducement to men to join.[68] The main effect is to make men's work easier, although as will be argued in Chapter 7, this may also create great problems for women and even reduce overall production levels.

Various factors have contributed to the official emphasis on introducing cash crops through men, perhaps the most important of these in many areas being their under-employment relative to women. Michael Moore, among others, has stressed the great reduction in men's traditional activities in Africa and Asia as a result of colonial rule: war, hunting, and in many cases pastoral stock-rearing were eliminated or much reduced by colonial policies, white settlement and increased population pressures.[69] Scarlett Epstein points out the importance in South Asian villages of the large amounts of leisure time available to men: they spend hours in the village cafés talking to each other and to outsiders, as important contacts are found, news passed on, gossip exchanged, and deals made. Over 50 per cent of the customers are from outside the village. Some of the local men may spend at least two hours a day there, while no village women ever go at all; they would be regarded as 'lazy' if they were to do so.[70] A similar situation also may be observed in villages in many parts of Africa.

The dual economy

In reaction to the men's virtual monopolization of many areas of innovation in agriculture, women have often tried to strengthen their control of the subsistence sector, and in some areas of petty trading as well. Achola Pala emphasizes that this, far from being a retreat to traditional women's work, has meant increasing numbers of women doing what was traditionally men's work in agriculture.[71] Ingrid Palmer sees women as having increasingly dominated the subsistence farming sector in Africa as men were forced into prolonged absences at colonial plantations.[72] In a dry-land and therefore undeveloped village in South India, Scarlett Epstein observed that women continued to cultivate the family plots, thus enabling the men to take wage-earning opportunities elsewhere.[73]

Women's efforts to build up the subsistence sector as a counter-weight to men's cash-cropping activities can be quite frantic. Uma Lele talks of the 'inordinate' amount of time spent by women on their food crops; their insistence on giving these priority over working on the export cash crops controlled by men is seen as a major obstacle to the development projects, which almost invariably attempt to promote the cash crops.[74] John de Wilde makes a similar observation:

> 'The fact that the man in much of Africa keeps the income from cash crops usually causes the woman, who has responsibility for feeding the family, to insist all the more tenaciously on producing all the family's food needs.'[75]

Despite strenuous and often ingenious efforts to adapt to the polarization of agricultural activities, the women's subsistence sector frequently declines in areas of agricultural commercialization. Food price policies, favouring the politically important urban population, often have a detrimental effect on efforts to produce a surplus of staple food items for sale.[76] SIDA points out that lending institutions do not deal with women, largely because credit is based on land titles or in certain cases, especially in co-operatives, on the coming year's yield from the perennial cash crops.[77] Marketing opportunities designed to aid small producers often discriminate against the low-priority basic foods: in a review of co-operatives throughout the world, it was found that subsistence crops are rarely included in activities.[78] Women in many areas have taken up petty trading on their own account, and are particularly prominent in this activity in much of West Africa, to the extent that it is frequently assumed that women often control a disproportionate share of the wealth there. Leith Mullings, who has reviewed the quite extensive literature on this question, concludes that this is in fact an illusion: although there are a few very wealthy traders, the overwhelming majority are in a poor position to earn more than a marginal subsistence despite their very strong market associations; the work is precarious and very time-consuming; and the women are tending to lose ground to the formal marketing sector.[79]

The situation as regards agricultural development policy in Botswana is an example of what can happen to women in an increasingly divided economy. The major export product is cattle, while subsistence is increasingly based on crops. Cattle belong overwhelmingly to men, who also do most of the rather low level of work involved;[80] women do virtually all the work in crop production, as well as in small stock such as pigs and poultry.[81] Cattle were the focus of the only development programs undertaken by the British colonial government before independence; climate and conditions are favourable, and the local stock good. As a major income-earner, especially for the richer people and including many politicians, cattle-rearing is seen as the only agricultural activity of any interest.[82] What little extension for crop production is available, moreover, is concentrated on men. The approximately 50 per cent of rural households without a man therefore find themselves totally excluded. Contacts are focused almost entirely on a small élite group of men who have some technical equipment and on those in the Pupil Farmer Scheme; virtually the only women participating are the widows of men previously enrolled. Extension meetings in the villages, moreover, usually take place in the *kgotla*, which is the men's traditional meeting place; the women, ironically, are usually unable to attend because they are at the lands, some distance away, where the actual crop production takes place.[83]

The concentration of inputs on promoting men's cash crops has implications also for land tenure. John de Wilde notes that, in Africa, in areas where land is held on a customary law basis, the sale of the land has started in response either to extreme land scarcity, or to the emergence of commercialized agriculture.[84] Ester Boserup cites survey data from Tanzania which provide an example of this process: a village close to 'modern' activities centred around a large plantation showed a much higher level of male ownership, mainly through purchase, than areas further away where women retained a greater degree of control, and a third, located between the two, showed an intermediate stage.[85]

The concentration of land ownership in male hands through the commercialization of agriculture parallels the increasing concentration on a small minority of more favourably placed landowners, when really large-scale inputs are involved as in the case of India's 'green revolution'. A dependency relationship is created on an even greater scale, since only the rich landowners are in a position to invest the necessary resources in cash crop production, while subsistence cultivators play safe with staple crops only.[86] Particularly hard hit by the introduction of high yielding variety (HYV) packages have been the pulses, a vital protein supplement in poor people's diet, which could no longer be inter-cropped with wheat as in the traditional system prevailing in some areas. Government policies, being strongly biased to the development of HYVs, helped to depress prices and production levels.[87] At the same time, the process of concentration on the HYV crops has eliminated many of the activities carried out by women, referred to by Michael Moore as 'interstitial' activities because they relied on products either growing wild or intercropped with subsistence food crops. On the Gangetic plain, for example, almost all land has now been converted to permanent tilling based on imported fertilizers and other inputs. The encroachment on forests and waste-land has resulted in the destruction of large parts of the 'ecological matrix' on which so many of women's economic activities depended, such as tending poultry and livestock, straw-plaiting and weaving, or collecting wild plants and spices for home consumption or sale.[88]

The same process has been observed in Africa: in Mozambique, for example, Sherilynn Young reports that the Portuguese colonial government in the 1950s and 1960s tried to compensate for the stagnating production of coffee, a crop which was interplanted between the staple bananas, by introducing upgraded cattle and tea. Both enterprises claimed the open grassland between the villages, which had been an important resource for women in producing the protein-rich bambara and groundnut products.[89]

Disadvantaged as they are when identified with the subsistence sector of a dual economy, once this has been taken over by the demands for land or labour from the men's cash sector, women may be in a much worse position. In a North Indian village, Rampur, which is located in an area of highly developed, irrigated and mechanized agriculture, women in Jat (peasant usually landowning caste) families have the worst of all worlds in respect of agricultural development. They now do extremely hard manual labour while the men do much less; one man commented, 'In this village, when it comes to work in the fields, we say 'ladies first'.'[90] At the same time:

> 'They have no control and little part in the decision-making about the family farm, in terms of what is to be grown and its marketing. Nor do they have any control over the money obtained from selling the crop. The men do all the management and the marketing and the women are not seen as having any automatic rights to the proceeds.'[91]

Conclusion

With a combination of professed ignorance and outright hostility towards women's rights to land, men in governments and in the development organizations have been systematically attacking women's links to the land in the process of legislation, and in the reform of land title in the Western mould. The LLDP case study indicates how the distorted methods of data collection and interpretation, outlined in Chapter 4, can lead directly to what amounts to the theft of women's land, and the systematic destruction of a way of life that could offer important advantages in organizing growth with distribution of the wealth created. Taking women's rights to land has enormous implications in terms of the incentives offered to them for their work on the land (discussed in more detail in Chapter 8), while at the same time the allocation of absolute ownership of land to a few men can lead to the exact opposite of what is intended: absentee ownership, and sometimes the abandonment of productive land because rights over it are no longer conditional on cultivating it. If agricultural production increases, it is at the expense of equity for women and their dependants and increasing duality in the economy and within every family. Too often, this kind of land 'reform' in fact leads to declining yields and contributes to the growing crisis for many Third World countries in national food self-sufficiency.

References

1. Harold Brookfield, *Interdependent development*, London, Methuen, 1976, p 3.
2. Ester Boserup, *Woman's role in economic development*, London, George Allen and Unwin, 1970, p 57.
3. Achola Pala, *African women in rural development research trends and priorities* (OLC Paper No 12), Washington DC, Overseas Liaison Committee, American Council on Education, 1976, pp 11-12, 14, 19, 20. On women's control of crops, see pp 6, 8-9, 26.
4. *Ibid*, pp 16-18.
5. *Ibid*, p 19.
6. David Edwards, *An economic study of small farming in Jamaica*, Jamaica, Institute of Social and Economic Research, University College of the West Indies, 1961, pp 95-97.
7. Lucy Mair, *Marriage*, Harmondsworth, Penguin, 1971, p 140.
8. Olivia Harris, *Women's labour and the household*, discussion paper for the BSA workshop on the peasantry, 13 March 1976, pp 13-14.
9. Lucy Mair, *op cit*, p 31.
10. Audrey Richards, *Land, labour and diet: an economic study of the Bemba tribe* (published for the International African Institute), London, Oxford University Press, 1939, p 190.
11. Vanessa Maher, 'Kin, clients and accomplices: relationships among women in Morocco,' in Diana Leonard Barker and Sheila Allen (eds), *Sexual divisions and society: process and change*, London, Tavistock

Publications, 1976, p 64.
12. Lucy Mair, *op cit*, p 63.
13. M G Swift, 'Capital, saving and credit in a Malay peasant economy,' in Raymond Firth and E S Yamey (eds), *Capital, saving and credit in peasant societies: studies from Asia, Oceania, the Caribbean and Latin America*, London, George Allen and Unwin, 1964, pp 137-38.
14. Polly Hill, *Migrant cocoa farmers of southern Ghana: a study in rural capitalism*, Cambridge, Cambridge University Press, 1963, p x.
15. Vanessa Maher, *op cit*, pp 65-66.
16. Achola Pala, *op cit*, p 20.
17. John de Wilde, *Experiences with agricultural development in tropical Africa*, Vol I: The Synthesis (published for the International Bank for Reconstruction and Development), Baltimore, The Johns Hopkins University Press, 1967, p 143.
18. *Ibid*, pp 132-33.
19. Ester Boserup, *op cit*, p 57.
20. George Dalton, 'Economics, economic development, and economic anthropology,' in Robert Heilbroner and Arthur Milord (eds), *Is economics relevant?* Pacific Palisades, Calif, Goodyear Publishing Co, 1971, pp 175-76.
21. R S Rattray, *Ashanti law and constitution*, London, Oxford University Press, 1929, p 23.
22. Claude Lévi-Strauss, *The elementary structure of kinship* (translated from the French by James Harle Bell, John Richard von Sturmer and Rodney Needham), London, Eyre and Spottiswoode, 1969, p 116.
23. Maurice Godelier, *Perspectives in Marxist anthropology*, translated from the French by Robert Brain, Cambridge, Cambridge University Press, Cambridge Studies in Social Anthropology 18, 1977, pp 105-6.
24. Mary Douglas, 'Is matriliny doomed in Africa?', in Mary Douglas and Phyllis M Kaberry (eds), *Man in Africa* (sic), London, Tavistock Publications, 1969, pp 121-23.
25. *Ibid*, pp 124-33.
26. Edith Clarke, 'Land tenure and the family in four selected communities in Jamaica,' *Social and Economic Studies*, Vol I, No 4, 1953.
27. Evelyn Reed, *Woman's evolution: from matriarchal clan to patriarchal family*, New York, Pathfinder Press, 1975, pp 165-66 and ff.
28. *Ibid*, p 331.
29. Polly Hill, *op cit*, p 16.
30. Mary Douglas, *op cit*, pp 132-33.
31. V W Turner, *Schism and continuity in an African society*, Manchester, Manchester University Press, 1957, p 113.
32. Interview in Abidjan, September 1977.
33. Institut Ivoirien d'Opinion Publique, *Sondage: droit rural*, Abidjan, IIOP, mimeo, 1975.
34. IIOP, Survey of Sodésucre, n d, p 101.
35. *Ibid*, p 104. Translation by the author.
36. John de Wilde, *op cit*, pp 140-41.
37. Interview with a senior UNDP official, Lilongwe, August 1977.
38. Interviews and discussion with LLDP personnel.
39. Bill H Kinsey, *Rural development in Malawi: a review of the Lilongwe Land Development Program* (Africa Rural Development Study Background Paper), Washington DC, IBRD, mimeo, 1974, p 1.

40. Interviews and discussion with LLDP personnel.

41. For example, an important indicator is whether a respondent agrees with the position: 'If you make a lot of money it is right that you should share it with your extended family.'

42. 'We have to eliminate their ideas and customs.' Interview in Lilongwe, August 1977.

43. Sources for pp 131-138 are confidential.

44. Interviews with LLDP personnel.

45. Ester Boserup, *op cit*, pp 60-61. On the Rhodesian case, see also John de Wilde, *op cit*, pp 138-40.

46. John de Wilde, *op cit*, p 137.

47. *Ibid*, pp 143-44.

48. *Ibid*, p 142.

49. Achola Pala, *op cit*, p 3.

50. D R Reynolds, *An appraisal of rural women in Tanzania*, Washington DC, Regional Economic Development Services Office, US Agency for International Development, mimeo, 1975, p 17.

51. Erich H Jacoby, in collaboration with Charlotte F Jacoby, *Man and land: the fundamental issue in development*, London, Andre Deutsch, 1971, pp 320-22.

52. Margaret R Haswell, *The economics of development in village India*, London, Routledge and Kegan Paul, 1967, Chapter 1.

53. Guy Hunter, *The administration of agricultural development: lessons from India*, London, Oxford University Press, 1970, p 20.

54. Swasti Mitter, *Peasant movements in West Bengal: their impact on agrarian class relations since 1967*, Cambridge, Department of Land Economy, University of Cambridge, Occasional Paper No 8, 1977, p 17.

55. B H Farmer, *Pioneer peasant colonization in Ceylon*, London, Oxford University Press, 1957, pp 207, 290.

56. M G Swift, 'Capital, saving and credit in a Malay peasant economy,' in Raymond Firth and E S Yamey (eds), *op cit*, pp 147-48.

57. André Gunder Frank, *Capitalism and underdevelopment in Latin America: historical studies of Chile and Brazil*, New York, Monthly Review Press, Modern Reader Paperback, 1969, p 28.

58. Patricia M Garrett, *Some structural constraints on the agricultural activities of women: the Chilean hacienda*, Madison, Wisc, Land Tenure Centre, Research Paper No 70, 1976, p 28.

59. *Ibid*, pp 1, 36.

60. Vanessa Griffen (ed), *Women speak out! A Report of the Pacific Women's Conference*, Suva, Pacific Women's Conference, 1976, p 51.

61. William A Allan, *The African Husbandman*, New York, Barnes and Noble, 1965, p 346.

62. Sherilynn Young, *Fertility and famine: women's agricultural history in Southern Mozambique*, Working paper, mimeo, n d, p 13.

63. Ester Boserup, *op cit*, p 54.

64. Cited in Margaret Haswell, *The nature of poverty*, London, Macmillan, 1975, p 34.

65. *Ibid*, pp 85-87.

66. D R Reynolds, *op cit*, p 10.

67. Swedish International Development Authority (SIDA), *Women in developing countries: case studies of six countries*, Stockholm, SIDA, 1974, p 31.

68. Agricultural Development Council, *Role of rural women in development*, New York, Agricultural Development Council, Seminar report on the Conference on the role of rural women in development, Princeton NJ, 2-4 December 1974, held by the Agricultural Development Council under its Research and Training Network Program, 1975, pp 35-36.

69. M P Moore, *Some economic aspects of women's work and status in the rural areas of Africa and Asia*, Brighton, Institute of Development Studies, IDS Discussion Paper No 43, 1974, p 19.

70. T Scarlett Epstein, *South India: yesterday, today and tomorrow: Mysore villages revisited*, London, Macmillan, 1973, pp 112-13.

71. Achola Pala, *op cit*, p 22.

72. Ingrid Palmer, *Monitoring women's conditions*, Working paper, mimeo, n d, pp 16-17.

73. T Scarlett Epstein, *op cit*, p 43.

74. Uma J Lele, *The design of rural development: lessons from Africa* (World Bank Research Publication), Baltimore, The Johns Hopkins University Press, 1975, pp 27-32.

75. John de Wilde, *op cit*, p 22.

76. See Uma J Lele, *op cit*, pp 31-32, 45.

77. SIDA, *op cit*, p 33.

78. Orlando Fals Borda, 'The crisis of rural co-ops: rural co-operative problems in Africa, Asia and Latin America,' in J Nash, N Hopkins and J Dandler (eds), *Co-operatives, collectives and co-participation in industry: popular participation in development*, The Hague, Mouton, n d.

79. Leith Mullings, 'Women and economic change in Africa,' in Nancy J Hafkin and Edna G Bay (eds), *Women in Africa: studies in social and economic change*, Stanford, Calif, Stanford University Press, 1976, pp 248-53.

80. Interviews by the author with officials of the Ministry of Agriculture, Gaborone, September 1977. See also Carol A Bond, *Women's involvement in agriculture in Botswana*, Gaborone, Ministry of Agriculture, 1976; Malcolm J Odell, Jr (ed), *Report on the sociological survey of the Losilakgokong area, Kweneng District*, Gaborone, Rural Sociology Unit, Division of Planning and Statistics, Ministry of Agriculture, Rural Sociology Report Series No 2, 1977; Carol Kerven, *Report on Tsamaya village, North East District*, Gaborone, Rural Sociology Unit, Division of Planning and Statistics, Ministry of Agriculture, Rural Sociology Report Series No 1, 1976.

81. Carol A Bond, *op cit, passim*.

82. Interview with senior official of the Ministry of Agriculture, Gaborone, September 1977.

83. Carol A Bond, *op cit*, pp 42-50. Interview with officials of the Ministry of Agriculture.

84. John de Wilde, *op cit*, pp 133-34.

85. Ester Boserup, *op cit*, p 59.

86. T Scarlett Epstein, *op cit*, pp 92-143 and ff. See also Margaret R Haswell, *Economics of development in village India*, London, Routledge and Kegan Paul, 1967, p 81. Also Michael Lipton and Mick Moore, *The methodology of village studies in less developed countries*, Brighton, Institute of Development Studies, IDS Discussion Paper No

10, 1972, p 60.
87. Ingrid Palmer, *Food and the new agricultural technology*, Geneva, UNRISD, 1972, pp 58-64.
88. M P Moore, *op cit*, pp 21-23, 31 and ff.
89. Sherilynn Young, *op cit*, pp 136-37.
90. Monica das Gupta, *'Ladies first'*, draft of paper for the 4th World Congress of Rural Sociology, Poland, mimeo, August 1976, p 4.
91. *Ibid*, p 5.

Chapter Seven

Women's work: its economic importance

Women's subsistence work

Enormous efforts are made by development researchers, and considerable sums of money spent, to discover exactly how much time is spent in field work: preparing the ground, sowing, weeding, pest control, transplanting, harvesting and associated operations. The focus of attention here is, of course, the men, consistent with the persistent bias in data collection observed in Chapter 4. Women are relegated to the unpaid 'family labour' category which is itself very inadequately measured, if at all. 'Non-farm' activities are given virtually no attention, and this is consistent with the observation, often made, that men in rural areas do remarkably little productive work other than field work, and that the non-farm work they do can easily be put aside, if necessary, since it is not essential to daily subsistence. This approach is singularly inappropriate to women. If there is one broad generalization that one can make about rural women, it is that their 'non-farm' work is strenuous, takes enormous amounts of time, and is absolutely essential to the survival of the family concerned. The division between farm and non-farm activities is a very unhelpful one in terms of measuring the contribution to subsistence made by women and men respectively, particularly with regard to the most basic need of all: food production.

In order to produce food in edible form, an enormous amount of work is involved after the harvest — which is the point where the official measurement of 'work' usually halts. Threshing, winnowing, drying, boiling (especially for rice paddy), and other activities have to be undertaken between harvesting and storage of many staple foods. Correct storage is also a major concern, since post-harvest losses of food can make all the difference between survival and destitution. As the food is used, exhausting work is again involved in processing it before cooking; for grains there is stamping, for example, and for cassava (also known as manioc) a lengthy series of processes to extract the poisonous substances. Before cooking, an adequate supply of water and fuel needs to be obtained, and these two requirements can take up to several hours a day. Water is of course a basic need in its own right, and has to be available for drinking in fairly large quantities, especially in hot, dry weather. Fuel may be obtained in the form of firewood, dried animal dung or crop residues, and in far too many countries is becoming increasingly scarce and difficult to find. Water, fuel, and especially wood, are very heavy and strenuous effort is required to carry these to the place where food preparation is done. Finally, of course, there is the cooking, which with many staple foods takes a considerable time especially if fuel supplies need to be conserved. Various relishes, spices and vegetables are needed to make the food palatable, and these have either to be gathered from the wild or cultivated as a separate crop which is also often excluded from the statistics on agricultural production. In many cases also pigs, poultry and other livestock are kept, and have to be watered and cared for. Apart from the preparation of food for home consumption, which constitutes the major part of subsistence agriculture, it is very important that any surplus production be marketed in order to produce some cash for the innumerable demands made on a family: taxes, school fees, clothing, food in times of shortage, and other basic needs. All of these essential activities performed by women — post-harvest work on the crops, storage, food processing, provision of water and fuel, cooking, provision of relishes, care of small livestock and marketing — tend to be relegated by development researchers to the despised 'domestic' sector and therefore disregarded by all except the home economists.

According to the circumstances, different aspects of the provision of food can take up a disproportionate amount of time and energy. Providing water is one example. The most detailed information available on this refers to parts of East Africa; one estimate is that water-carrying there accounts for up to 12 per cent of the day-time calorie usage of most of those involved, and up to 27 per cent in very dry or very steep areas. A separate study of nine villages in different parts of Tanzania found that the water requirements of the average household needed more than four hours of adult water-carrying time per day, and over an hour of child time; the work was done almost entirely by the women (as seems generally to be the case now, except where water is sold for cash). In a study of rural Delhi, India, women were found to be spending an average of almost an hour a day fetching water.[1] Given the enormous amount of work involved, it is obvious that nearby dirty water will often be used in preference to more distant clean water. Since it is obtained at such cost, it will also be reserved for drinking and cooking rather than non-essential washing. Watering of animals and gardens will also be very limited by the extra effort involved.

The problem of obtaining fuel is one with multiple implications for the subsistence base of an area. Fuel, like water, is indispensable to the provision of edible food, and the need for it grows directly in line with population growth and the need for food. In many areas where there is intense population pressure on the available land, forests have been cut down to provide more crop area. The demand for firewood has meant increasing scarcity of trees and bushes in many places, greatly increasing women's labour input in seeking out and carrying home, over long distances, enough wood to cook a few meals. Since it has been estimated that the average user burns a ton of firewood a year,[2] when wood becomes scarce the labour involved in this task adds up to a considerable expenditure of time and energy. There are other implications: deforestation, and sometimes the removal of all twigs and residues from the ground, leaves many areas open to rapid soil erosion and therefore a reduction in food crops. Deforestation of watersheds leads to disastrous flooding which kills people and their livestock, and also devastates crops. In the attempt to replace firewood, many people, especially in Asia, are drying animal dung for use as a fuel, which robs the soil of vitally important nutrients and organic matter.[3] Yet, because the provision of fuel is seen as part of the 'domestic' and therefore irrelevant sector, the problem has been virtually ignored by many development planners. They have forgotten that, as one Indian official protested, 'Even if we somehow grow enough food for our people in the year 2000, how in the world will they cook it?'[4]

Another of women's tasks which takes up large amounts of time and energy is food processing; this includes the grinding, husking and pounding of grains and other basic foods, as well as the brewing of beer from these foods for home consumption and, increasingly, for sale as the major source of cash for many women. Food processing can often be an important source of cash even where the work is invisible to most development researchers: for example, secluded Moslem women in many parts of Nigeria process food for sale in their own homes.[5] The same is true of many parts of Asia.

Phyllis Kaberry, describing the intolerable workload of Bamenda women in the colonial Cameroons, refers to food processing as 'the last straw that breaks a woman's back after a long day's work on the farm and a weary trudge home.[6] A number of studies have been carried out in Africa on the time required to prepare maize by manual methods; in one case it took 13 hours just to pound enough maize to feed a family for between four and five days. Winnowing and soaking take additional time even before cooking can start.[7] According to estimates made in the former Congo, processing of tapioca and maize took four times as long as all the work hours spent on the cultivation of these crops.[8] Obviously, this is a major constraint on food production; one report speaks of African women 'sitting about hungry with millet in their granaries and relish to be found in the bush' because they were too exhausted to tackle the heavy, three-hour work of preparing the food for eating.[9]

The problem may be particularly acute in the more sparsely populated continent of Africa, since grinding mills are not readily available to many rural women; however, it is by no means confined to Africa. In Mexico at one time, the grinding of grain by hand took four to six hours every day per family where there were no mills available. In fact the women's need for mills to relieve the work of grinding led to their successfully demanding the reopening of a mill which the men had decided to close down.[10] In India, a study of adult women from cultivating families in a village near Delhi, concluded that the average time spent on milling alone was 1.3 hours per day. A similar study in rural Java, which did not disaggregate activities to the same extent, showed an average of 3.2 hours per day per adult women spent on food preparation.[11] The need for grain mills is great, and women will often devote whatever cash they can to relieving the heavy work of food processing. Another strategy is to switch to other kinds of crops that need less processing, or alternatively to buy already processed foods such as imported wheat flour or rice; this has considerable consequences for a country's self-sufficiency in basic foods, but can also have serious consequences for the nutrition available to poor people.

The total subsistence work-load

The total amount of subsistence work done by women is not recorded in most studies of the labour force, or in 'manpower' studies and planning. Time-budget studies are just beginning to appear rather sporadically which do cover some or all aspects of women's work. With overwhelming uniformity, they depict rural women as working extremely long hours, and expending energy without adequate rest on a wide variety of tasks, all of which are essential to a family's survival. In most if not all cases, women's work is seen as much more arduous and time-consuming than men's. In a discussion of the problem of 'unemployment' in Zambia, a recent mission of the International Labour Organisation (ILO) showed that women's overwork was at least as important as men's under- or unemployment. In fact women in urban areas suffer much more severely from the lack of paid jobs than men, while:

> '. . . in the rural areas it seems widely true that women are over-worked rather than under-worked, in the sense that they work very long hours, are pressed by many duties and obligations, are responsible for much of the work in agricultural production as well as for virtually all the food preparation, housework and the care of children . . .'[12]

Very little detailed time-budget study involving all activities has been done for African women, however. A survey of women's work in resettlement villages of Upper Volta describes them as working a 15-hour day starting at five in the morning, during the growing season, and mentions that the physical demands of the work leave them exhausted. The worst tasks are seen as preparing the millet by pounding and grinding, and fetching water — neither mills nor wells having been provided by the project management since this work is seen as irrelevant to agricultural production.[13] A similar observation at a rice re-settlement scheme in Kenya indicates that about 11 hours a day were involved in various activities, six days a week, at all times of the year, although the classification of different kinds of work is perhaps more limited than the Upper Volta study.[14]

The most detailed time-budget work has been done in various Asian countries. The study of a rural Delhi village, already mentioned, records adult women as spending an average of 4.9 hours a day on 'domestic' tasks, including grain-milling, fetching water and fuel, cooking and the rest, and 4.5 hours on agricultural tasks: a total of 9.4 hours a day. However, child-care (including the breast-feeding of babies, obviously a necessity for subsistence) is specifically excluded from this total.[15] Other observations from North India indicate that men are able to take a day off on Sundays, and have very slack periods at certain times of the year, whereas women's work continues with little variation at a high level.[16] Several case-studies of rural women in a wheat-growing area of Haryana have shown that the average working day for women, if all work is included, is between 15 and 16.5 hours. In households with only one adult woman, she would be so overworked that she would scarcely have time to nurse her baby. Old women also worked extremely hard; one 75-year-old woman was still labouring 10 hours a day preparing food for sale. Men, on the other hand, had a much less strenuous work-load, and plenty of time to smoke and play cards.[17]

A time allocation study in rural Bangladesh by Abdullah Farouk showed that women were spending between 10 and 14 hours a day in productive work (defined to include income-generating and expenditure-saving work), as opposed to between 10 and 11 hours for men.[18] More detailed studies have been made of time allocation by women and men in rural Java, Indonesia. For women aged 30 and over, 5.5 hours are spent every day on 'domestic' tasks, most of this on food preparation, and 6.7 hours on income-earning tasks, a total of 12.2 hours a day. For all women aged 15 and over, the total number of hours worked averaged 11.1, as compared to only 8.7 for the men.[19]

Several studies are under way at the time of writing to provide better time budgets for rural women, and to improve the criteria for describing activity as 'work'. The studies mentioned here tend to have relatively complete criteria, in comparison with others that arbitrarily assign women's non-farm work to the category of 'housework' or even 'leisure'; however, even the best of the studies available so far are particularly inconsistent in regard to cooking and child-care, presumably because these two categories are the closest approximations to Western ideas of domestic work, or non-work. In terms of their importance to a family's subsistence, however, these two tasks cannot be so arbitrarily dismissed: without cooking, staple foods are almost all inedible, while the category of 'child-care' covers the breast-feeding of babies, feeding and care of children, including medical care, and training and supervising children in their own contribution to the household's work. It would appear that in many if not all areas, if their work is properly measured, it will be seen that women are spending over 10 hours a day in essential activities relating to family subsistence, especially in the provision of food. This extremely strenuous work may involve up to 16 or more hours a day, in other words, constant labour with hardly any leisure time or adequate rest. Women are being expected to keep up this kind of pace for an entire life-time, with little relief in old age, in the face of often inadequate nutrition and with the severe physical demands of frequent pregnancies, childbirth and lactation.

Women's farm work: a major determinant of food production?

The various kinds of non-farm work carried out by women are clearly essential to subsistence in terms of providing water, edible food and other basic items. Lack of energy and time with which to carry out the various tasks will often mean short-cuts which have a serious impact on the amount and quality of food and water available; in other words, women's energy and time are important constraints on the delivery of basic needs. While the focus here has been on food and water, which in fact take up most of the energy and time available, there are other subsistence tasks carried out by women. These include house-building and repair, the production and sale of handicrafts and other manufactured goods, marketing, care of the sick, and of course the care and training of children — which are essential to health, welfare and the maintenance of the social structures on which a family's subsistence is based. Not only is women's labour input a major determinant of production in non-farm subsistence work, but there is some evidence that it is increasingly a constraint on the production of subsistence crops in terms of the field work itself. In some studies, women's labour input has been shown to be the critical constraint on crop production.

Africa

In Africa, according to Uma Lele, labour availability is a constraint on agricultural productivity to a greater extent than elsewhere, particularly Asia, where land availability is a more important factor.[20] It is in Africa, too, that women's agricultural work has been recognized by some observers: Lele for example mentions the 'disproportionate' effort by women in agriculture, which can be double the amount of field work done by the men even if all non-farm work is excluded.[21] And since men's non-farm work is frequently less critical to subsistence than the women's, the male labour input is more elastic than that of the female. This creates a serious labour bottleneck, especially at peak seasons, for women's labour in the fields, which together with the non-farm work can create intolerable demands.[22] The men's and women's labour bottlenecks occur at different times, particularly given the men's refusal to do many of the tasks to which little value is attached, like weeding — tasks currently allocated mainly to women.[23] It is very often weeding that is the crucial bottleneck, since it has to be done within a very specific period of time in order to provide a reasonable yield from the crop in question. Generally speaking, it may be women's crops and women's fields which are the most affected by the impossible demands made on women's time and energy at peak periods. An official report on rice production in Gambia, for example, stresses the 'seasonal limitations' involved in this women's crop as the major constraint on production.[24] In many cases the women's crops will be the staple food crops. In Ghana, for example, it is estimated that almost all the food crops are cultivated, harvested and marketed only by women.[25] In Botswana, women are increasingly responsible for all crops.[26] The pattern seems to be common to much of Africa.

A series of surveys of labour inputs into agriculture in Zambia have indicated that women's labour is in fact the major factor in production there. It was found that differences in the number of males in a farming family did not influence the number of acres sown (in an area where land shortage was not itself an important constraint); however, the number of females available was the most important variable in determining acreage:

> 'The tables suggest that the presence of women in the family is more acreage-inducive than the presence of men, oxen or even tractors (with the exception of maize farmers in Mumbwa where the effect is roughly the same). The tables are further evidence, therefore, of the key role played by women in the acreage-decision . . .'[27]

The crucial operation, it is emphasized, is weeding, which is done — as is the majority of the cultivation — by women.[28] Women work an average of 42 per cent more than men in the fields, and the degree of skill used in cultivation, as well as the energy and time expended, are particularly critical, given the use of improved seeds which can only provide a better yield with a high standard of husbandry, for example planting in rows with the correct spacing, and adequate fertilization of the soil.[29] The importance of women's work is even greater in the families of 'farmers' using more innovative methods than those of 'villagers' which are more traditional.[30] The report is marred, unfortunately, by the observation of the men in charge — contrary to their own findings about women's key role in the production process — that it is the presence of women as *consumers* that 'exerts the principal pressure on the [male] cultivator to increase his acreage'.[31] The researchers in question allowed their prejudices about women to obscure the extremely important findings of their research about women's key role in agricultural production. This refusal to recognize the obvious is of course widespread. A welcome exception is the observation by the ILO mission to Zambia:

> 'The general conclusion which emerges from surveys and interviews is that women are in general over-worked in rural areas, that women's labour is one of the factors which determine how much land can be cultivated and how well, and that *the pressure on women's time is an important constraint on raising agricultural production and rural living standards*.'[32] (Emphasis added.)

Asia

If in Africa the problem is to get planners to recognize women's work as an important constraint on agricultural production, in Asia there is the task of persuading planners and researchers that women do any work at all. A book on Pakistan claims that in rural areas, 'a woman's life is household activities, gossip, grooming, visits to female relatives or shrines, and religious festivals'.[33] Innumerable text books on rural development in the sub-continent describe agricultural processes as if the work was done only by men.[34] However, one of the very few studies to seek information on women's farm work, compiled by questioning the men in the North-West Frontier Province of Pakistan, shows that women are doing a wide range of essential work. This involves particularly the post-harvest processing of grains and the maintenance of livestock. Their work was probably underestimated 'as [male] farmers do not readily volunteer information about women or their families'.[35] The farm work reported, in an area where relatively strict *purdah* was common although not universal, was a mixture of activities performed in the home compound plus those performed in the fields. Work reported as done by women in over half the households included cleaning seed, drying grains, feeding livestock and milking dairy animals. Processing the milk is invariably done by women, including pasteurization, fermenting for yogurt, and producing butter. Women are also responsible for determining the surplus quantity of grain production which is to be marketed; they build the grain storage bins, and are responsible for the sun-drying of grain. In general, there is no difference in the participation by women in farm-work and food processing inside the compound, according to ethnic group and the degree of *purdah* practised. A difference did show up in the reporting of women's field work, where about one-third of the men said that the women of their families were involved, almost all of them from the more flexible communities.[36]

It is significant that outside observers would simply not see the work being done by most of the women, since it is inside the compound where male strangers are not admitted; but being literally invisible to men does not render this work any the less vital for family subsistence. The problem of subsistence work being done by women out of sight of male researchers is one which is common to many countries, such as those in the Middle East and North Africa, where *purdah* is observed to some degree. What is surprising is that the existence or otherwise of *purdah* seems to make relatively little difference in terms of the measurement of women's work; this is almost always ignored regardless of whether or not it can be seen. As long as the attempt to assess the importance of women's work is limited to the questioning of their male relatives by male researchers, the question of the degree to which women's work is a constraint on agricultural production, will remain unanswered.

Some data is beginning to appear on India and Bangladesh. In Bangladesh, where there is a considerable degree of *purdah*, one study has stressed that women's work in food production is very important, particularly in fruit and vegetables which form an important part of the diet; their work is seen as directly related to the variety as well as the quantity of food available to the family. Women also process the staple paddy after the harvest.[37] A few women are involved in fish-breeding and fishing, while others produce important medicinal herbs.[38]

Policy-makers in Bangladesh, urban élite men (and some of the élite women) often take the view that poor women are dependants who sit at home and are idle.[39] Germain stresses that not only do Bangladeshi women work extremely hard at a variety of productive activities, including the production of rice, livestock and fruit and vegetables, but that they are being drawn increasingly into new field work, including irrigation and work on new crops. Since the war of liberation, large numbers of women have become the sole support of their families.[40] In addition to subsistence production, the poorer women do labouring jobs for others in the area and also very hard manual work on a variety of Food For Work projects. In addition to the amount of energy and time involved in women's work, there is the very important factor of their skills and expertise even in new techniques for which they have had no formal training. Even strict *purdah* is no barrier to their familiarity with the details of field work. One study emphasizes the importance of women's expertise and their participation in decisions about crops:

'Those who have worked with rural women here know that, far from being ignorant, they are, like their husbands, highly skilled and knowledgeable about the work that they do. And they understand how it relates to the work their husbands or sons do. In recent interviews with rural women, we found that although some, especially younger women, could not say exactly where their husband's land is because they could not get out to see it, they know what crops were planted, when they were planted, what the yield was, and in the case of HYV [high-yielding variety] rice, how much of each kind of input was needed and how much it cost. In response to questions about those aspects of farm work that women are responsible for, they explained how they knew when rice is properly parboiled, how they can tell when it is properly dried, how and when they prepare the courtyard for drying the rice, what special steps they take in processing and storing seeds, what causes breakage and loss of rice in processing. Although they seem to have received little modern knowledge about the aspects of rural work they are engaged in, they are resourceful and expert within the limitations imposed on them.'[41]

In a particularly favoured area of Northern India, where new technologies of rice production have revolutionized agriculture from a subsistence to a cash-cropping operation, Monica Das Gupta found that women are deeply involved in the new techniques. The area is one of the wealthiest in India, due largely to successful agriculture and the tending of carefully bred cattle. Most of the work involved in both these activities is done by women, and they are actually doing more of the work now than previously.[42] The reasons for this are discussed in more detail below. A study of another wheat-growing area in India, Haryana, concluded that women's contribution to crop production alone was always more than 50 per cent, while their work in animal husbandry and 'farm support activities' at home would make their percentage contribution to farm production even higher than that.[43]

In a very different environment, the Philippines, Gelia Castillo has found that women are crucial to the farm production process and to decisions on the use of available resources; yet development planners have failed to allow for their participation:

'It is relatively rare for females to be included in rice and corn production training despite the fact that much of the labour input in production is contributed by females. Women are often responsible for the care of pigs and chickens in the backyard but they are not recognized as livestock raisers. Although it is a well-known fact that Filipino women participate actively in decisions affecting the farm, and are almost always in charge of marketing farm products, they have never been a deliberate target clientele for agricultural development programs. Furthermore, the wife is the acknowledged treasurer of the family so credit, savings, and investment programs might be profitably addressed to them as much as to the men even if to counteract her veto.[44]

In a detailed study of women's work in a rural area of Java, Indonesia, Ann Stoler observed that this work was crucial to rice production, the basic crop, not only in terms of hours worked but also in terms of the intensity and timing of periods of peak activity. The total labour inputs per hectare are higher for women than men.[45] The female activities of planting and harvesting also need to be done quickly and at a precise time. Harvesting in particular, the most labour-intensive of all agricultural activities, demands large supplies of labour at concentrated periods of time. For a small field, the preparation of which by men requires 20 days' work over a period of one month (i e one man working three to six hours a day), planting requires the same amount of work-time concentrated into one morning (i e twenty women at a time) and harvesting about twice as much for the same size of field.[46] Most of the women, particularly in the poorer households, support the family by hiring themselves out for harvesting and other agricultural work. They also provide the most regular portion of household cash income through various activities outside rice-agriculture, including production and sale of coconut, sugar and other food items, as well as doing work for wealthier women.[47] The importance of women's contribution to family subsistence is recognized by the fact that, during harvest time, men will take over much of the women's non-farm work such as child-care and cooking.[48] This study makes an observation that may be generally true of rural women: that it is in the poorest households that women make the greatest relative contribution to subsistence, and where their contribution is most explicitly recognized by all members of the family.

Latin America

Carmen Diana Deere, in surveying women's work on the Latin American *minifundios* (smallholdings), has mentioned a number of practices which obscure women's crucial work in agriculture. One of these is the servile relationship of smallholders with the big landowner, which prevails in much of the Andean region, whereby the labour services of the whole family are appropriated by the landlord for work on his own *latifundio*, in return for their usufruct rights to a small parcel of land. Whole families may also be hired by a landowner, especially at harvest time, with no money paid directly to the women for their work. There are also systems of customary labour reciprocity which include collective work groups — involving women, men and children — as well as the exchange of labour services for products and the exchange of labour between families. Women play a major part in these activities, but because no cash changes hands they are not recorded as significant to production.[49] Changes in the definitions used in agricultural labour censuses have also resulted in an increasing distortion of the record; in 1950, 43 per cent of all women were counted as 'economically active' as compared to only 22 per cent in 1960, and the proportion of those who were reported to be involved in agricultural work was similarly reduced.[50] Deere suggests that, in fact, woman's importance as agriculturalists has increased considerably as poverty has intensified and men have started to take other jobs wherever possible:

> 'The concentration of landholdings, combined with demographic growth, have required male semi-proletarianization in the rural areas with the woman remaining as the primary agriculturalist on the subsistence land plot.'[51]

In a case study of Northern Peru, Deere found that the declining trends in Peruvian census data for women in agriculture reflected the opposite of the true situation. Since it was arbitrarily considered — by statisticians but also by many of the respondents — that any man resident in a farming family was automatically the major agriculturalist, women's participation was largely suppressed. This problem was compounded by the fact that there was a time limitation as regards seasonal work that applied only to 'unremunerated family workers', mainly the women, and that they were therefore not even being recorded as auxiliaries even though their labour was crucial at peak seasons such as the harvest.[52] A rather more complete result is presented in the 1976 Peasant Family Survey, which estimates the proportion of agricultural work done by women. However, this suffers from another arbitrary exclusion: it eliminates from the data collection 'intermittant participation' of less than four hours at a time, as well as complementary agricultural activities such as preparing food for the farm labourers, collecting manure, tool repair and animal care.[53] Deere observes that this very seriously understates women's contribution:

> 'Most commonly, a peasant woman will combine her productive activities over the length of the working day, alternating her work

in the fields with meal preparation. But if a large noon day meal is required, which is generally the case when non-family labour is employed in agricultural work, a woman must dedicate more of her time to cooking than to field work . . . The number and quality of meals is an important component of the remuneration to labour when wage labour is employed.'[54]

Having excluded much, perhaps most, of women's agricultural work, the survey then produces estimates of their contribution which amount to 21.4 per cent of total agricultural work, and 25.4 per cent of the work on the *minifundio* itself.[55] The degree of underestimation is indicated by the fact that whereas only 38 per cent of the households reported initially that women participated in agricultural production, data on actual participation, through a detailed investigation for age and sex in each individual task of the agricultural cycle, showed that in fact, the correct figure should have been 86 per cent.[56] Over 90 per cent of all the women sampled said they usually weed by hand, an important task in the cultivation of grains; they hand-pick various crops, husk corn, collect grain during threshing, plant seed and shake soil free from the roots of weeds while planting. Seventy per cent of the women participate in breaking ground with picks, hoeing, and reaping grain. A smaller proportion take part in ploughing, threshing grain and carrying in the crop from the fields.[57]

As in Java and elsewhere, there is a major difference in women's work according to the family's income level and resources; however, at none of these levels does the amount of agricultural work fall, it simply changes in character. Women in households which employ others in their fields spend much of their time preparing the labourers' meals. The poorer the household, the more crucial is women's participation in the actual field work: whereas women from landless families participate in 73.6 per cent of the 15 field tasks listed in the Peasant Survey, women from middle and 'rich' peasant households participate in only 61.9 per cent, although this is still a high figure. The greatest differences are in the tasks normally done by men; for example, the only women who admit to ploughing are those from smallholder households with no adult male and no money to hire one.[58] Deere concludes:

'Women's greatest agricultural participation, relative to men, is found among the poorest strata of the peasantry, those without sufficient access to land to produce their full subsistence requirements.'[59]
'As the family loses access to the means of production of subsistence, the importance of agriculture in generating familial subsistence

also declines . . . As the relative importance of agriculture diminishes, agriculture appears less as a male occupation and more a familial activity. Not only do all family members contribute their labor time to agricultural production, but decisions concerning agriculture are shared to a greater extent, or fall totally to the woman, as another concern to add to her domestic responsibilities.'[60]

Where have all the young men gone?

In the Peruvian case, women are increasingly responsible for farm work as well as their other subsistence activities because of the reduction in male labour inputs, especially in the poorest families. Poor women are much more likely to participate in what are generally regarded as male tasks. This change in the customary division of labour is the result of men being absent from the farm to take paid jobs wherever possible.[61] Deere has suggested that this phenomenon is widespread in Latin America, and is increasing as subsistence agriculture becomes less viable as a family's economic base.[62]

It is interesting that the reverse happens, to some degree, in the case of Java: as already noted, men take over women's subsistence work because the women have better opportunities for earning cash at various times.[63] Since the trend is increasingly toward the greater availability and remuneration of jobs for men, however, coupled with the increasing concern about male 'unemployment' already noted, there is a growing identification of the subsistence farming sector as women's, along the lines of the Latin American model. This is of women invariably having to take over men's subsistence work, never the other way around. In fact, since women's work is given little or no value and dismissed as domestic work or 'housework', men may consider it beneath them to do women's work, or what has become labelled as such. In much of the Third World, it is now accepted that the Western model of women being destined for 'domestic' work and the care of children is the norm — particularly in the media and official interpretations of society. Data is sadly lacking on this, but it may be that the primary impact of Western influence on gender-roles in a society is to transfer much or all of the non-farm subsistence work to women, followed later by a transfer of subsistence farming.

The division of labour by gender is, as explained in Chapter 1, very flexible. Levine, in his analysis of African societies from which there is a high rate of labour migration, found that the division of labour by gender is often modified but at the same time accentuated. The men become mobile, and women are more bound than before to the homestead because they have to try and do the men's work in addition to their own. They do not gain any more rights to cash from the sale of produce but in fact lose ground in some cases.[64] Jane Wills has data from Kenya and Uganda to show how a 'shift' in the division of labour has been occurring, where women do many tasks previously considered for men only, but with men very rarely taking over women's work. She also observes that men retain or intensify their control over the sale of produce.[65] Uma Lele makes the general observation that in Africa the division of labour, although 'bewilderingly diverse', is a major factor in restricting labour availability (from the men), because of their refusal to do the despised 'women's work'.[66] The logical consequence of such trends is that as women take over men's tasks in agriculture, and men maintain a rigid refusal to do 'women's work', the subsistence farm work done by men will grow less as time goes on.

The process probably originated with colonial occupation and, in Africa, with the mass slave trade: men were forced out of their homes by kidnapping, forced labour or the demand for taxes in cash which could be obtained only by work on plantations or mines. In addition, the deliberate introduction of new cash crops to men diverted their energies away from subsistence agriculture. Some of men's jobs were also eliminated by colonial policies, such as the ban on hunting wild game, or the removal of people to overcrowded 'reserves' in order to free the best land for European settlement, which upset many systems of shifting agriculture in which men cleared and prepared the land. In some cases these moves also brought people into the tsetse-fly belt, eliminating men's work in cattle-herding.[67] All of these processes helped to build up a situation which justified the extraction of 'unemployed' men from the subsistence economy and their use in low-wage jobs for the benefit of the colonial metropole, or local business interests set up by colonial settlers.

Men's greater mobility, both potential and actual, has probably operated in a very complex way to increase the burden of subsistence work for women. Pepe Roberts has offered a very interesting description of the process in Niger, where both the size and the composition of family labour has altered considerably in recent years. The decline in household size is mainly characterized by the break-up of family control over male labour; the older men in charge of the extended family have much less control over their younger male relatives since they are able to leave the area for paid employment. 'Male access to cash incomes through migration, conflicting with the household head's responsibility for providing for the household and paying taxes on behalf of all adult members, is partly the cause of this.' Since the younger men are absent for part of the year, it is preferable for the elders to facilitate their entry into independent farming and into taking responsibility for their own taxes. However, they have retained control over the women, who cannot leave so easily, and have clearly pressurized women to take over tasks previously categorized as being male. The women resent this fiercely, invoking the authority of the religious leaders, the marabous, against their doing men's work in addition to their own, when the resulting production is even more tightly controlled by the male elders than before — their rights to land from which to make a small income of their own having been reduced to a 'concession'. Men freely admit that they now give women the worst plots of all.[68] The loss of control over young men, leading to increased pressure on the women, may be a widespread phenomenon. Margaret Haswell noticed something rather similar in Gambia, where the breakdown of the extended family into nuclear family units had undermined the system for pooling male labour and so made it more difficult to clear the land by the axe and fire for millet production. This made it necessary for people to rely more heavily on the rice crop, which had traditionally been the sole responsibility of women, and led to great hostility between women and men, the women reacting by trying to force the men to assist with their rice crop.[69] In many areas of shifting cultivation, the clearing of land by burning piles of timber is essential to the production of nutrients for crops, which otherwise would have soils too poor to support them. In Zambia, for example, where this system is known as *chitemene*, the break-up of matrilineal clans together with migration of young men to the mines or elsewhere has been an important factor in undermining the entire basis for food production.[70]

The increasing loss of control over male labour for subsistence, which then becomes identified, almost in its entirety, as women's work, is associated not only with the physical absence of men from the subsistence household, as they migrate temporarily or permanently to work elsewhere, but also with their refusal to work even when present in the household. They may become consumers instead of producers for the subsistence unit. In her North Indian study, Monica Das Gupta observed that in families with several women, the men would prefer to stay at home rather than help with harvesting: 'the men take a

holiday'.[71] In Kenya, only 54 per cent of males over 17 regularly work their own holdings, as compared with 85 per cent of the females.[72] In Botswana, men have started to demand cash payment for ploughing with their cattle, instead of doing this as part of reciprocal family obligations as previously; women tend to continue working for each other in turn, or for payment in kind.[73] As husbands and sons are often absent, leaving women to produce crops alone, many of them are completely without male labour and cattle — usually owned by men — to do essential soil preparation, unless they can afford to pay.[74] Some women span the cattle themselves, with the help of children or a neighbour, but this is likely to be less effective than the old system.[75]

The absence of younger men, or their refusal to work under the old system of obligations to the extended family, means a sudden rise in the dependency ratio for the family concerned; there are the same number of dependants — children, old people, the sick and handicapped — being supported by fewer able-bodied adults. If the men actually consume more of the household production than they generate, as may well be the case, they actually add to the number of dependants to be supported by the women. At the same time, of course, many so-called dependants, notably children, are in fact producers in their own right. However, just as the value of children's work is becoming a recognized factor, it is being rapidly eroded in many areas by the spread of education, particularly for boys who are given priority. Not only does school and sometimes homework take up much of the children's time, but the content of the teaching makes them contemptuous of subsistence farming and the boys, in particular, may refuse to do any, as I was told by villagers in Ivory Coast and elsewhere. Women can often recoup some help only by keeping their daughters at home at peak work periods, which of course further handicaps the girls in terms of their access to higher education and to paid employment.

The increasing loss of men's and boys' labour from subsistence, and their tendency to consume more than they produce, may be the last straw for the precarious balance maintained in the poorer households, and the capacity of the women to keep the system going. Margaret Haswell observed a syndrome in Gambia, which may well apply fairly generally, where men are sporadically present but not doing much subsistence work. The periodic visits of the lower-paid men to the village are often not at the time when their labour is needed, at the harvest period,

and when there is plenty of food. These visits occur more frequently in the 'hungry season' when they are feeling economic
stress and decide to live off their families for a while.[76] Remit-
tances from male migrant workers are often negligible, not com-
pensating for the loss of their labour, and it tends to be spent
on men's entertainment and consumer goods rather than on
basic family subsistence.

Women cannot leave their increasingly intolerable situation in
the rural areas as readily as men; paid employment is very much
harder to find, their ties to their children are often much closer,
and their families may have much more control over them than
over young men. In many cases, if they leave the area they also,
unlike men, abandon all family ties and all rights to be supported by their families as well as their right to land. In fact growing
numbers of women are leaving the land for the towns, particularly in Latin America. This perhaps represents the last stage of
decline in smallholder subsistence, a state of desperation reflected in the number of women working as prostitutes or servants
in the cities.

Development planning: making things worse?

A recognition of women's work as essential to subsistence, and
as an increasingly critical factor in agricultural production, is
almost completely lacking from the work of development
planners. They suffer the myopia of labelling women's subsistence work 'domestic' and therefore to be dismissed as trivial.
James Brain describes a fairly typical case of misunderstanding
among Tanzanian officials and overseas aid 'experts' with regard
to the constraints on women's energy and time; they 'could not
see why the women were unable to carry out an eight-hour
working day in the fields, and assumed that demands on
women's labour input could be made almost without any
limit'.[77] Just as damaging as the assumption that women's
labour availability is limitless is the idea that it is completely
unavailable. A common assumption in Bangladesh, for example,
is that rural women will not work outside their homesteads; this
is directly contradicted by the fact that experienced program
staff know about poor women's work outside, including field
work and labouring jobs in Food For Work programs.[78] This
would seem to be a clear example of the kind of contradictions
which can, according to situational analysis, be an integral part
of an ideology about gender roles, and echoes some of the conversations I have had with officials (outlined in Chapter 3)

where they see women working but do not recognize this as relevant to their work in planning. The reliance on quantitative techniques can also have an important effect in ensuring that reality is not recognized by planners; Pepe Roberts mentions that in Niger, the omission of any reference to the cost of labour in the cost-benefit analyses of the new inputs, is linked to the assumption that all labour will be supplied by women ('family labour') to which no value or 'opportunity cost' is attached.[79]

Many attempts to increase agricultural production rely on new or intensified labour inputs for tasks performed partly or wholly by women. For example, the new Japanese method of rice production, which is being introduced to increase output in many Asian and other countries, uses much more labour than current methods. This is especially the case in planting and transplanting seedlings and in weeding; and women play a vital part in these activities.[80] A similar trend is evident in the introduction of High Yielding Varieties (HYV) for various crops. The introduction of HYVs in India, in the so-called 'green revolution', has probably exacerbated the imbalance in demand for female and male labour, especially in weeding — a major job for women. In terms of casual labour, HYVs increase the need for women's labour from 53 days per acre to 63, while that of men fell slightly from 19 to 18.[81] In addition, HYVs of rice require more work at various stages after the harvest, which adds significantly to women's workload.[82] In a North Indian area, the 'green revolution' expanded women's work in agriculture, particularly with the increased number of crops grown per year. Another factor is that, since there is no longer any grazing land left fallow, the women have to carry green fodder daily to the cattle from the fields.[83] In Malawi, where different packages of innovations are involved, a number of practices designed to raise crop yields also require greatly increased labour inputs from women, both because of the rigid new division of labour, already mentioned, and because there are so few able-bodied men available for the new tasks of pre-rain hoeing and ridging, and the adherence to strict weeding schedules. The package has proved 'very difficult to apply'.[84]

In the Philippines, it has been reported that 'family labour' has been inadequate in some areas to cope with weeding, which has been estimated to account for some 35 per cent of the labour involved in the cultivation and harvesting of HYV rice. This is at least double the use of labour required by previous varieties of rice. However, HYVs generally were observed to lead to labour-saving in land preparation, which was widely mechanized, and labour-creating in straight-row planting, weeding and other jobs.[85] In Africa, the heavy emphasis on monocropping as opposed to mixed cropping has considerably increased the weed problem, as has the use of fertilizer, which, on poor soils, can lead to the appearance of weeds not previously known; mechanization of land preparation can also create an overwhelming weed problem.[86] In general, new open-field systems on the Western model, used in tropical environments where intercropping and shifting agriculture had been the norm, can no longer prove viable because of the new weeding problem created.[87]

As Jane Wills points out, the extra work imposed on women by planned innovations in agriculture is not relieved by any labour-saving inputs for many of the operations:

'Often the only type of innovations possible for the women farmers are those involving the use of [their own] additional labour, and not those involving the reduction of labour due to mechanization.'[88]

The reverse is often true for men's work, where enormous expenditure and infrastructure are devoted to the provision of mechanized land clearance, and ploughing in particular, and sometimes also to harvesting and threshing. The increasing use of tractors, for example in India, has greatly reduced men's work in agriculture.[89] Scarlett Epstein has suggested that the substitution of capital and machinery for labour, in India, is a reaction to the very unstable nature of the male labour force, with men oscillating between village and town.[90] A vicious circle would seem to be in operation; men's labour is unreliable, therefore it is reduced to the lowest level possible, which in turn encourages rural men to abandon their agricultural work.

Mechanization of men's work only, can cause innumerable problems in the absence of comparable labour-saving for women's tasks. John de Wilde notes that it increases hostility between women and men, and raises general tensions within the family.[91] The use of tractors for men's work can be seen in many cases to be justified, not only by the economics of agricultural production, but as an instrument in men's attempt to reduce their commitment to agriculture. De Wilde mentions that in much of Africa, men tend to use any implements or machinery, regardless of cost, to achieve the same output targets with less labour input from them. He cites a case in Uganda where a tractor-ploughing service was used to reduce men's work, with almost no evidence that there was any increase in production as a result.[92] Uma Lele comments that in Tanzania, tractors are often used as an inducement to men to form ujamaa villages, even though the machines are of no great value to the production process.[93] The mechanization of men's work thus makes no sense in removing a labour bottleneck, since it is the women who are overworked. It is their weeding and other tasks which constitute the immediate constraint on production; the enthusiasm for tractors, or any kind of innovation in land preparation, can be explained more convincingly in terms of men's withdrawal from agricultural work. Men do not, however, abandon their claims to control the operation; in fact the access to additional inputs like tractors tends to increase their control of the whole family's labour.

One major problem is that, where labour-saving technologies are introduced which apply to women's work, they have been handed over to male control. Small implements such as presses, grinders or cutters have been given or sold on credit to men by development agencies, even when the work for which they are a substitute is traditionally done by women. For example, corn grinders have been made available in Kenya but women are not taught to operate them. Oil presses in Nigeria, tortilla-making machines in Mexico, and sago-processing machines in Sarawak are also purchased and operated by men, partly because only they have access to cash or credit. There is a very high demand for food-processing machinery from women, but without control of the equipment they are able to relieve the pressure of work only by continuing reliance on men's machines, which involves spending whatever cash they can accumulate for this purpose. The situation helps to reinforce the stereotype that women cannot manage machines, and that they can cope only with the most low-productivity and low-value operations.[94]

Control of women's work is concentrated more and more in the hands of a few men. Another phenomenon is the application of a new technology to work which had previously provided a relatively high return to women's labour, such as the hand-harvesting in Java already noted as a major source of family income. Several farmers and landlords have begun to limit the number of women employed in harvesting their fields. Alternatively they reduce the wages given them, particularly by the employment of a small number of men, using sickles, to whom they pay a cash wage and thus replace larger numbers of women, who harvest with the small-bladed *ani-ani* and are paid a share of what they harvest.[95] A similar process is that of the replacement of women's hand-pounding of the rice-harvest, in which the returns approach those of harvesting, by the introduction of hulling-machines.[96] Both these jobs, since they offer a relatively high return for labour, are of great importance to women as alternatives to the usual low-productivity tasks which provide negligible returns for back-breaking work. Needless to say, there is great opposition among the local villagers to these practices, but it seems that they are unable to prevent them since the landlords are increasingly able to concentrate wealth, and therefore control, of the agricultural process in their own hands.[97] John Harriss has characterized the unequal distribution of mechanization for female and male labour as being an important factor in upsetting social relationships of interdependence — the system of checks and balances between the needs of landowners for labourers and of labourers for farmers.[98] At the very least, the impact of innovations on women's and men's work respectively needs to be carefully considered in the planning process. At the moment, it seems that planners are failing to take this into account, with serious consequences for the distribution of wealth by class as well as by gender.

Life and health

With women's work being so basic to subsistence for rural families, their health is a matter of vital importance affecting directly their capacity for work. Even more crucial is their survival rate, particularly in mid-life when they already have children and must provide for them even where men have abandoned some or all of their responsibilities. Yet there is every indication that the increasing overwork of rural women, and the growing dependency burden, is undermining their health and strength and, in many cases, in combination with the dangers of pregnancy, childbirth (and the alternative, abortion) leading to premature death. In many countries there is a peak in deaths among women of working age; in Nigeria, for instance, a working life table shows that mortality among women aged 15-64 accounts for 85 per cent of all losses from the labour force.[99] Very little concern is shown by planners about these losses because women's work is not valued, and women's lives — as noted in the case of cost-benefit analysis — are therefore not seen as valuable.

There are many stages of chronic debilitation, the magnitude of which is even greater than that of women's deaths but which is hard to quantify; very little attempt has been made to measure the effect of chronic overwork on malnourished people. In terms of the impact of planned innovation on women, however, it is important to note that where their labour is increased, even when it is paid, this is often about half of what is paid to male labourers. The wages paid (including food in Food For Work projects) may well be inadequate for women's basic calorie requirements (quite apart from the need for specific nutrients), let alone the needs of their children and other dependants. Women who are pregnant or lactating, or who are doing very heavy work (as would be the case in terms of their double load of subsistence work as well as wage labour) have the same or greater calorie requirements than most men. It is often noted, however, that men have the first choice of the family's food, and they have most of the available cash with which to buy extra food and also drinks — frequently alcoholic — which are made from staple foods that would otherwise be consumed by the entire family.

The whole question of women's health and strength is a crucial one to which much more attention needs to be paid. Suffice it to say that much development planning has an adverse effect by depriving women of important resources and by intensifying their work-load, while reducing their control over their own work-patterns. In cases where the only way to increase the help available for subsistence work is to bear more children, or where contraception is not available, pregnancy and childbirth bring enormous risks to life and health which, in combination with a history of malnutrition and overwork, can overwhelm the woman involved. It is a physical impossibility to expend more calories in work than one takes in as food, except at the expense of body tissue; pregnancy and particularly lactation likewise place enormous physical strains on a woman's body. The long-term deprivation of women is related to the extremely high infant death rates in developing countries: in Gambia, for example, where 'one found women collapsing by the roadside on the way back from the [rice] swamps and having to be helped home, the cause apparently being too much work with too little food,'[100] nearly 40 per cent of the babies borne by these women died within four years of birth. An article in the *British Medical Journal* listed infectious disease and inadequate maternal care as being probably the major factors contributing to high mortality and the impaired growth of small children, associated with the strenuous agricultural work of the adult woman, who is 'expected not only to bear and to rear children but also to produce a main food crop, rice, almost unaided'.[101] This was the result, as already noted, of men leaving the subsistence-based extended family which had made it impossible to organize land-clearing and so meant almost total reliance on the women's rice crop.

Hostility

In the development process, the division of labour is increasingly a matter of men abandoning their traditional obligations and women being forced to take over their work in the struggle for subsistence. Control of the family is becoming concentrated in the hands of men because they have so much more time, mobility, education, access to land and other resources, particularly the most important of all: cash. Increasingly, there are indications that as families become involved in development, the men are showing contempt for women and a more directly hostile and oppressive attitude.

> 'When asked why wheat should not be ground in the flour mill to spare the women from the arduous job of grinding wheat, Buro's son replied, "It costs money." When he was told that a family owning a big farm could afford it, he replied half in jest and half in earnest: "This keeps them physically fit; moreover an idle brain will be the devil's workshop." This type of reply was received from most of the men interviewed.'[102]

'The attitude is that if this one goes, another can always be got.'[103] Both observations come from the heart of the 'green revolution' area of North India, from the men who have benefitted the most from the process of development.

References

1. Cited in Michael P Moore, *Some economic aspects of women's work and status in the rural areas of Africa and Asia*, Brighton, Institute of Development Studies Discussion Paper No 43, March 1974, p 14.
2. Keith Openshaw, 'Wood fuels the developing world', *New Scientist*, Vol 61 No 883, 31 January 1974.
3. For an overview of the problem, see Erik P Eckholm, *The other energy crisis: firewood* (Worldwatch Paper 1), Washington DC, Worldwatch Institute, 1975.
4. Quoted in *ibid*, p 12.
5. *Role of rural women in development*, Seminar report of the conference on the role of rural women in development, Princeton NJ, 2-4 December, 1974, held by the Agricultural Development Council under its Research and Training Network Program, October 1975, p 1.
6. Ann Oakley, *Sex, gender and society*, London, Temple Smith, 1972,

p 138.

7. For a general summary see J H Cleave, *African farmers: labour use in the development of smallholder agriculture*, New York, Praeger, 1974, pp 215-68; also Michael Lipton and Mick Moore, *The methodology of village studies in less developed countries*, Brighton, Institute of Development Studies, Discussion Paper No 10, June 1972, p 55, n 4.

8. Ester Boserup, *Woman's role in economic development*, London, George Allen and Unwin, 1970, p 57.

9. *Ibid*, p 165.

10. *Ibid*, p 164.

11. Quoted in Michael Moore, *op cit*, pp 13-14.

12. International Labour Organization, *Narrowing the gaps: planning for basic needs and productive employment in Zambia*, Report to the Government of Zambia by a JASPA (Jobs and Skills Program for Africa) Employment Advisory Mission, Addis Ababa, ILO, January 1977, p 47.

13. Société Africaine d'Etudes et de Développement (SAED), *Etudes sur les besoins des femmes dans les villages de l'AVV et proposition d'un programme d'intervention*, Ouagadougou, Ministère du Développement Rural, Aménagement des Vallées des Volta, 1977, p 36.

14. Jane Hanger and Jon Moris, 'Women and the household economy', in Robert Chambers and Jon Morris (eds), *Mwea: an irrigated rice settlement in Kenya* (Ifo-Institut für Wirtschaftsforschung, Afrika Studien 83), München, Weltforum Verlag, 1973, p 226.

15. Quoted in Michael Moore, *op cit*, p 15.

16. Monica das Gupta, *'Ladies first'*, draft of paper to be presented at the 4th World Congress of rural sociology, Poland, August 1976, p 3.

17. Shantri Chakravorty, 'Farm women labour: waste and exploitation', *The Journal of Social Change* (Delhi), March-June 1975, Vol 5, Nos 1 and 2, pp 9-11.

18. A Farouk and M Ali, *The hardworking poor*, Dacca, Dacca University Bureau of Economic Research, 1975.

19. Benjamin White, 'Population, involution and employment in rural Java', in Gary E Hansen (ed), *Agricultural development in Indonesia*, Ithaca, NY, Cornell University Press, 1976.

20. Uma J Lele, *The design of rural development: lessons from Africa* (World Bank Research Publication), Baltimore, The John Hopkins University Press, 1975, pp 23 and ff.

21. *Ibid*, pp 26-27.

22. *Ibid*, p 25; see also J H Cleave, *op cit*, p 180.

23. Uma Lele, *op cit*, pp 26-27; also John de Wilde, 'Vol 1: The Synthesis', *Experiences with agricultural development in tropical Africa*, (published for the International Bank for Reconstruction and Development), Baltimore, The Johns Hopkins University Press, 1967, pp 84-85.

24. Margaret Haswell, *The nature of poverty*, London, Macmillan, 1975, p 143.

25. Polly Hill, *Migrant cocoa farmers of southern Ghana: a study in rural capitalism*, Cambridge, Cambridge University Press, 1963, p 164.

26. Carol A Bond, *Women's involvement in agriculture in Botswana*, Gaborone, Ministry of Agriculture, 1974.

27. Universities of Nottingham and Zambia Agricultural Labour Produc-

tivity Investigation (UNZALPI), *Some determinants of agricultural labour productivity in Zambia* (Report No 3), Lusaka, University of Zambia, November 1970, pp 16, 20.

28. *Ibid*, pp 20, 36 and *passim*.
29. *Ibid*, pp 21, 38.
30. *Ibid*, p 16.
31. *Ibid*.
32. ILO, *op cit*, p 47.
33. Quoted in Nici Nelson, *Why have women of village India been neglected? An examination of the literature on the role of women in India's rural development*, mimeo, Institute of Development Studies, p 2.
34. *Ibid*, pp 3-4.
35. Malik Ashraf, 'Notes on the role of rural Pakistani women in farming in the Northwest Frontier Province', *Land Tenure Center Newsletter* No 55, January-March 1977, p 10.
36. *Ibid*, pp 10-13.
37. Women for Women Research and Study Group, *Women for women: Bangladesh 1975*, Dacca, University Press, 1975, pp 46 and ff.
38. *Ibid*, p 62.
39. Adrienne Germain, *Women's roles in Bangladesh development: a program assessment*, Dacca, Ford Foundation, 1976, p 2.
40. *Ibid*, pp 2-3, 12.
41. T Abdullah and S Zeidenstein, *Rural women and development*, paper presented at the seminar on the role of women in socio-economic development in Bangladesh, Bangladesh Economic Association, May 1976, mimeo.
42. Monica das Gupta, *op cit*.
43. Shantri Chakravorty, *op cit*.
44. Gelia T Castillo, *On 'liberating' Filipino women: which women?*, mimeo, n d.
45. Ann Stoler, *Land, labor and female autonomy in a Javanese village*, unpublished paper, February 1975, p 19.
46. *Ibid*, p 20.
47. *Ibid*, pp 8-9, 35 and *passim*.
48. *Ibid*, p 23.
49. Carmen Diana Deere, *The division of labour by sex in agriculture: peasant women's subsistence production on the minifundios*, paper for the Department of Agricultural Economics, University of California, Berkeley, January 1975, pp 10-11.
50. *Ibid*, p 10.
51. *Ibid*, p 32.
52. Carmen Diana Deere, *The agricultural division of labour by sex: myths, facts and contradictions in the northern Peruvian Sierra*, paper presented to the panel on 'Women: the new marginals in the development process', Joint National Meeting of the Latin American Studies Association and the African Studies Association, Houston, Texas, 2-5 November 1977, mimeo, pp 6, 9.
53. *Ibid*, p 11 n6.
54. *Ibid*, p 15.
55. *Ibid*, p 11.
56. *Ibid*, p 10.
57. *Ibid*, pp 16-17.

58. *Ibid*, p 17.
59. *Ibid*, p 13.
60. *Ibid*, p 25.
61. *Ibid*, pp 17, 18, 21.
62. Carmen Diana Deere, *Agricultural division of labour, op cit, passim.*
63. Ann Stoler, *op cit, passim.*
64. R A Levine, 'Sex roles and economic change in Africa', *Ethnology* (Pittsburgh), Vol IV, No 2, April 1966, p 188.
65. Jane Wills, 'Peasant farming and the theory of the firm', in *Rural Development Papers* (Kampala) No 68, 1968; and 'A study of time allocation by rural women and their place in decision-making: preliminary findings from Embu District', *ibid*, No 44, 1967.
66. Uma Lele, *op cit*, pp 26-27.
67. See Ester Boserup, *op cit*, pp 53-55 and *passim.*
68. Pepe Roberts, *Critique of rural modernization programs in the Republic of Niger*, Paper presented at the 4th National Development Research Conference, Glasgow, 26-29 September 1978, pp 11-12.
69. Margaret R Haswell, *The changing pattern of economic activity in a Gambia village*, London, Her Majesty's Stationery Office, 1963, p 73.
70. See e g C J Trapnell and J N Clothier, *The soils, vegetation and agriculture of North Eastern Rhodesia*, Lusaka, Government Printer, 1953; and W Allan, *Studies in African land usage*, Manchester, Rhodes-Livingston Paper 15, 1949.
71. Monica das Gupta, *op cit*, p 4.
72. 'Women in Kenya', *Social Perspectives* (Central Bureau of Statistics, Ministry of Finance and Planning, Nairobi), Vol 3, No 3, April 1978, p 6.
73. Carol Bond, *op cit*, p 21.
74. *Ibid.*
75. Carol Kerven, *Report on Tsamaya village*, Gaborone, Ministry of Agriculture Rural Sociology Unit, 1976, p 4.
76. Margaret Haswell, *op cit*, p 206.
77. James L Brain, 'Less than second-class', in Nancy J Hafkin and Edna G Bay (eds), *Women in Africa: studies in social and economic change*, Stanford, Stanford University Press, 1976, pp 275-77.
78. Adrienne Germain, *Women's agricultural work in Bangladesh*, New York, mimeo, 1977, p 6.
79. Pepe Roberts, *op cit*, p 12.
80. Margaret R Haswell, *Economics of development in village India* London, Routledge & Kegan Paul, 1967, pp 70-73.
81. John Harriss, *Implications of the introduction of HYVs for social relationships at the village level*, Paper for a seminar of the Project on Agrarian Change in Rice-Growing Areas of Tamil Nadu and Sri Lanka, St John's College, Cambridge, 9-16 December 1974, mimeo, pp 6-11.
82. Women for Women, *op cit*, pp 66 and ff.
83. Monica das Gupta, *op cit*, p 3.
84. John de Wilde, *op cit*, pp 140-41.
85. Ingrid Palmer, *The new rice in the Philippines*, Geneva, United Nations Research Institute for Social Development (UNRISD), Studies on the 'Green Revolution' No 10, 1975, pp 148-53.
86. John de Wilde, *op cit*, pp 98-100; J E Mansfield, *Analysis of data from sample farms and co-operatives in Northern and Luapula Provinces,*

Zambia, Surbiton, Surrey, Ministry of Overseas Development Land Resources Division, Supplementary Report 11, 1974, p 52.

87. Clifford Geertz, *Agricultural involution: the process of ecological change in Indonesia*, Berkeley, California University Press, 1963, Chapter 10.
88. Jane Wills, 'A study of time allocation', *op cit.*
89. See Monica das Gupta, *op cit*; also John Harriss, *op cit*, pp 6-11.
90. T Scarlett Epstein, *South India: yesterday, today and tomorrow: Mysore villages revisited*, London, Macmillan, 1973, pp 246 and ff.
91. John de Wilde, *op cit*, p 99 n4.
92. *Ibid*, p 97.
93. Uma Lele, *op cit*, pp 33-34.
94. Irene Tinker and Michele Bo Bramsen (eds), *Women and world development*, Washington DC, Overseas Development Council, 1976, p 27.
95. Benjamin White, *op cit.*
96. *Ibid.*
97. *Ibid*, n31.
98. John Harriss, *op cit*, p 20.
99. Madugba I Iro, 'The main features of a working life table of the female labour force in Nigeria, 1965' *Journal of the Royal Statistical Society*, Series A, Vol 139, Part 2, 1976, p 259.
100. D P Gamble, *Economic conditions in two Mandinka villages: Kerewan and Keneba*, London, Colonial Office, 1955, pp 108-12.
101. I A McGregor, W J Billewicz and A M Thomson, 'Growth and mortality in children in an African village', *British Medical Journal*, Vol 2, 1961, pp 1665-66.
102. Shantri Chakravorty, *op cit*, pp 10-11.
103. Monica das Gupta, *op cit*, p 4.

Chapter Eight

Incentives

As it becomes more readily accepted that so-called 'primitive' people, and peasants in general, are very far from being the stupid, conservative barrier to change which is how they have been described, it seems that this new awareness of peasants as highly rational decision-makers is still stopping at the last frontier: women. Western male stereotypes of women include the idea that they are 'irrational' or 'illogical' because they do not conform to men's expectations. This prejudices the attitudes of Western-trained planners towards women in the Third World. Virtually no attention has been paid to the question of providing adequate incentives for women to participate in planned change. Payment for work done by women is often made to their husbands; and even more universal is the habit of paying men only for crops produced by the joint labour of all members of the family. This helps to transform husbands into bosses and wives into servants, creating great friction and also undermining the co-operation between partners and the quality of the work done by them.

A few planners have, in fact, noticed that the diversion of incentives to husbands can undermine a whole program or project. In East Africa, economically successful production of pyrethrum, a women's crop, was halted because of the formation of a co-operative to market the crop to which only men could belong. The women simply lost their incentive to produce, and started to withdraw their labour.[1] A similar problem arose with the same crop in Papua New Guinea, where, despite considerable promotion efforts by the Government, production was not sustained after the initial trials. The women were doing over half the work involved, and a total of 83 per cent if their children were included, but they received only a small fraction of the price paid. In many cases the resulting hostility between women and men escalated into violence, with several women being seriously injured by their husbands. They retaliated, it would seem, by refusing to grow the crop, despite considerable efforts by the extension workers to provide new incentives for the men.[2] It may even be that men are less responsive to direct incentives than women, because of their different use of money. It is apparent, particularly in Africa, that men's labour input into agriculture is evident only where there are cash crops with relatively high returns.[3] Whereas women have a constant need of money, even in small amounts, to supplement their subsistence work in providing household needs, men tend to receive and spend large amounts of money at one time. This leads to the familiar but little-mentioned phenomenon of all work coming to a halt for a few days after pay-day on some projects, while the men get drunk; one report on Zambia puts it rather euphemistically:

> 'The temptation to celebrate the end of a hard season's work and the fact that money is not usually saved, leads often to reckless spending.'[4]

In discussion of another Zambian project, it was recognized that incentives aimed at men would be spent, not on household needs, but on consumer goods.[5] Given the shortages of worthwhile consumer goods in rural areas, and the fact that most are imported, such a policy would seem to be very short-sighted for any rational approach to national development. Household needs, on the other hand, on which women spend much of their cash, are readily available in most areas and their purchase will often benefit other women and their families. The same considerations apply to the effect of changing wage-rates and the prices and marketability of crops grown and marketed by women.

Non-financial criteria of value can also be applied to this question, in particular the expenditure of time, something many rural women lack as the previous chapter has made clear. Study of the use of time can be made to analyze economic choice,[6] particularly where lack of time is a serious constraint on production. This explains why women would devote more of their valuable time to subsistence production, which they control, rather than on the men's cash crops. As Uma Lele observes, 'labour availability in smallholder agriculture is . . . closely related to the desire of subsistence producers to ensure domestic food needs'.[7] She complains of the 'inordinate' amount of time spent by women on food crops, which from a planners' point of view would seem irrational.[8]

Women's resistance to any increased expenditure of their time can be crucial to the acceptance or otherwise of a particular innovation. A project by the British National Institute of Agricultural Engineering in Gambia revealed that systems of

primary cultivation resulted in increased weed growth, imposing greater demands on women's time. This was likely to lead to
the loss of the whole crop because women refused to weed
it. On another occasion the Institute introduced a new, cheaper
type of plough for building ridges of groundnuts; visually the
work done by this tool was very similar to the existing one but
the people were reluctant to adopt it. After much trouble it
was discovered that the weeds emerged through the new pattern
of ridge rather quicker than through the original type. Although
the difference was small the women objected to the extra work
involved for them and the men were therefore not able to
accept the innovation.[9] De Wilde mentions similar reactions
as being a barrier to the introduction of otherwise sound
practices.[10] The choice of cash crop can also be very much
affected by the premium on women's time at different seasons.
For example, in Nyanza Province, Kenya, a cotton scheme
organized by the colonial government failed because that crop
required labour inputs from women just at the time when they
were very busy with their food crops. Women gave precedence
to spending their time on their families' subsistence needs,
and the cotton crop failed.[11] On the other hand, if the returns
on the time spent are high enough women may divert their
labour to a new activity: in a 'green revolution' area of the
Philippines, for example, a survey showed that the increase in
chemical and mechanical weeding inputs had been accompanied
by an increase in the time spent in weeding.[12]

The problem of a lack of incentives for women, combined
with very heavy demands on their time, can be illustrated by
various development projects which involve a major commitment to change. These, in many cases, have attracted people
initially by the high incomes offered.[13]

Mwea

At the Mwea resettlement scheme in Kenya, based on irrigated
rice production, Jane Hanger and Jon Moris found that the real
contribution of women to the success of the household was
much greater than was recognized in terms of their official
social and legal position there. Four basic assumptions underlying the scheme were found to be wrong: first, that priority
should be given to raising tenant incomes, on the assumption
that this would generate the fastest rate of growth and almost
automatically increase the welfare of all concerned; second, that
the tenants, who were given no land for food crops, would eat

rice; third, that tenants would use income from the rice harvest to buy other kinds of food; and finally —

> '. . . the Settlement procedures treat the male head of household as if he were the principal labourer and decision-maker for the irrigated fields farmed under a tenancy agreement with him, whereas both traditionally and in present practice the women contribute the larger share of farm work within the cultivation system.'[14]

Many of the women found the physical work involved in rice cultivation especially arduous, being additional to their own work. At the same time, assumptions about income from rice sales reaching the women were largely unfounded, any money having to be begged from the men in small amounts, or obtained from black market sales. The work of procuring fuel for cooking was very hard, or had to be replaced by purchases, again using money begged from the husband. The living conditions at Mwea were also uncongenial and unhealthy. In a survey of 99 scheme households, 86 women mentioned lack of money, 85 famine, 88 fuel, and 92 health as being the particular problems of women. The scheme as a whole was perceived by most as being an unpleasant place to live and work, despite the fact that Government services were concentrated on it.[15] Women lost their right to enough land to grow food, and the shortage of land for this purpose was seen by them as the major difficulty of life on the scheme. It continued to be their responsibility to feed their families by their own efforts, using maize and beans rather than rice.

For the women, therefore, the unirrigated areas set aside informally for food plots were more important than the irrigated holdings to which the management of the scheme devoted so much attention; these areas are described as 'the only resources of their own not under the man's control'.[16] This could have had an impact on women's time allocation decisions, at a peak period when the urgent need to plant the food plots at the optimal time for the short rains (a critical factor in maize yields in the area) clashed with the heavy labour input required for weeding the rice crop, a job imposed on the women. Another option was for the women to earn their own money by working for other villagers.[17]

An additional factor running counter to official objectives is that, although women were not paid by their husbands out of the cash received from the project, 'a man cannot refuse his wife access to the paddy she has helped to harvest with her own hands'.[18] This was used, not as intended for household consumption, but for sale outside the project. Black market trading in rice, which probably reached high levels at various times despite strenuous efforts by the management to prevent it, was largely a woman's affair and an indication of their conflict of interest with the objectives of the scheme. It had apparently not been seen as a possible solution for the management to purchase rice from women as well as men.[19]

A more drastic response to the difficulties created by the scheme was for women to leave altogether, causing considerable disruption to the productive unit. The researchers say that 'many women have found Mwea an intolerable place to live'. There seemed to be a high rate of desertion of male tenants by their wives, who had no interests to keep them there; and it seemed to be hard to attract women to become wives of the tenants under the prevailing conditions. Hanger and Moris conclude:

> '. . . it is our contention that the unsatisfactory recognition of a woman's rights and needs within the Scheme remains one of the greatest weaknesses of the "Mwea system".'[20]

It is a weakness which has been observed to affect other schemes, where very similar problems arise from management's assumptions about women. These emerge from their own gender ideology, rather than discernible fact.

Tanzania

James Brain mentions land allocation to women as the principal problem in Tanzanian settlement schemes. Women who had traditionally held considerable rights to land had none at all in the settlements. All such rights were vested in the husbands, and all proceeds of the land and labour were handed over to them. Women could not hold onto their land outside the scheme, either, because their rights were contingent on cultivation which they were too far away to do, and custom forbade the hiring of labour for this purpose. This not only created injustice on a routine basis — which the majority of the men as well as the women recognized as a severe problem — but also produced great confusion and insecurity for women in their old age or widowhood.[21] At one scheme at Kingurungundwa, women had revolted against 'the imposed condition of virtual serfdom' and demanded their rights. The response of the Government was not encouraging, and instead of resolving the women's problem it closed down the scheme shortly afterwards.[22]

The Volta Valleys

A third scheme where major difficulties have been analyzed as stemming from the lack of incentives for women is the series of resettlement villages set up in Upper Volta by the Government's Autorité des Aménagements des Vallées des Volta (AVV). This is a pilot scheme for massive investments by external donors to follow the campaign against onchocerciasis (river-blindness) in the underpopulated river valleys. A study by the Société Africaine d'Etudes de Developpement (SAED) explained the dissatisfactions, primarily of women but also increasingly of young people generally:

> 'AVV is trying to promote family welfare by means of modernizing family enterprises, adopting a system of production which involves only the head of that enterprise: other members of the family (women and young people) are involved only as "workers".'[23]

Women had absolutely no land on which to grow their families' food, nor did they have any control over the income which was passed by the management to the husbands, either in cash or — unlike Mwea — in kind. The work, moreover, was exceptionally hard for the women, who provided the main labour inputs essential for participation in the scheme: 15 hours a day during the growing season from April to January. In addition to work on the cotton fields, subsistence tasks such as fetching water and processing millet were made much harder by the failure of managements to supply the promised wells and grain mills.[24] With the loss of their previous sources of income, particularly the sale of food at local markets, women were unable to feed their children: 'since women had no personal resources they are obliged to condemn their children to constant hunger'.[25] The result was similar to that at Mwea: complaints from the wives of tenants, a succession of departures by some women followed by their husbands, who could not cope with the work alone, and threats from many others to leave.[26]

In these projects, women are expected to provide a major part of the labour force while rights to land are taken away and all financial incentives withheld from them, and given exclusively to their husbands. The result is acute dissatisfaction, conflict and significant numbers of women leaving altogether. The effect on overall project targets is negative, and in the Tanzanian project resulted in complete closure.

Undiscovered potential

Since there are so few development programs or projects that have attempted to offer adequate incentives to women, their potential for increased production remains undiscovered. I have found one case, however, where this happened more or less by accident: the recently completed project for Small-scale Irrigated Horticulture Training and Development at Chapula, on the Zambian Copperbelt, run by the Zambian Government, UNDP and the Food and Agriculture Organization (FAO). Aimed at the production of irrigated vegetables for the growing urban markets there, the project was unsuccessful from the financial point of view, a major drawback being the inadequate number of hours worked, particularly at harvesting time.[27] Women had to walk long distances to get there, and left at midday despite protests, saying they were absolutely exhausted and needed rest and food. Very few men ever appeared in the fields.[28] Particular difficulty was encountered in getting the registered growers to weed their plots.[29] The Terminal Report comments on the need to 'increase the growers' personal involvement and sense of responsibility toward the scheme'.[30]

The factor distinguishing the Chapula project from others in a similar position is that, although originally only men were considered for the training that led to registration as a grower, and to allocation of irrigated land, the women successfully challenged that. Unwilling to accept the position of 'dependants' or unpaid 'family labour' when they were in fact doing most or all of the work, a women's deputation met with the expatriate management and successfully demanded the right to participate in their own right.[31] As the Sociological Survey quaintly describes it: 'Those ready to join the scheme include adult dependants who wish to live an independent life once given the opportunity to join the scheme'.[32] One woman was accepted initially and used as a test case. Fortunately, she was successful and graduated from the course at the top of her training group. This then paved the way for other women to apply and be accepted for training. By the sixth course, in 1973, half the participants were women: 'The course had an excellent start and may be considered as the keenest course at Chapula since its foundation'.[33] By 1974, 36 of the 141 growers, or just over a quarter, were women.

Growers' income records for the first half of 1975 show that women registered as growers earned an average of K51.93 after deductions for inputs; men earned an average of K23.35. Thus, female growers earned on average twice the income of male growers relying on their wives' 'family labour'.[34]

This result at Chapula is perhaps indicative of the importance of project managers reviewing the incentives available to women, and also the barriers to their full participation in a scheme, despite the fact that they do much of the work. Some additional evidence was found by a study financed by the US Agency for International Development: agricultural or rural development projects in Peru, Nigeria, Kenya and Lesotho were found to have benefitted significantly by the participation of women although, in several cases, their involvement had not been planned by the management but was the women's response to attempts at innovation, as with the Chapula project. There were also important spin-offs from women's taking responsibility for new crops or new techniques, in that skills were passed along to many people outside the scheme itself, and were applied to crops other than those with which the project itself was concerned.[35]

John de Wilde, writing over a decade earlier, had already discovered the importance of women's participation for the success of a project:

'In view of the role of women in agriculture it is important that women as well as men attend [Farmer Training Courses]. The success of Kenya's FTC's is in no small part due to the opportunity for training which they have given to women.'[36]

In East African 'progressive farmer' schemes, de Wilde found that women were significantly more adaptable to new opportunities than men, and constituted one of the 'progressive' groups who 'seem to provide the best potential for extension work'.

'. . . we found quite a few women in the category of progressive farmers; widows or women whose husbands were working elsewhere and thus not living at home. There was a large measure of agreement among the agricultural staff on the general proposition that they were often more receptive to advice and instruction than men, but that women in most cases lacked the opportunity or authority to apply advice. When they are widows, however, or their husbands leave them in charge of the farm (and of the expenditure of the income therefrom), they have both the incentive and the authority to apply improved practices.'[37]

An interesting observation comes also from Malawi, where Jonathan Kidd found that households with adult men present ('male-headed') would be motivated by incentives only at the higher levels of profitability; however, women alone ('female-headed' households) would increase their labour input more readily in response to lower or medium-level incentives:

> 'The model presented here implies that under quite plausible assumptions women farmers may have greater economic incentives to innovate than male headed households.'

The potential remains unrealized, however, because of the unavailability of training, extension, credit and other inputs to women.[38]

Almost none of the women's subsistence activities, farm and non-farm, have been systematically reviewed. Nor have they been subject to improved technology and even minor improvements can have an important impact. The provision of wells, better fuel supplies, improved tools for farm and non-farm work would all help to reduce their expenditure of time, while an adjustment in development policies to ensure that they have direct access to incentives for increased labour input in key sectors, is also likely to show improvements in output. Many Third World countries face a crisis in terms of the need to import staple food because of the stagnating or declining domestic food-production sector, and this represents an important and still increasing strain on the balance of payments and on national development objectives, especially for the poorest or 'least developed' countries. In most of these, agriculture accounts for over 40 per cent of total production (even according to the conventional system of national accounting which understates the subsistence sector). Out of 14 countries which recorded negative growth rates of food production per head of population in 1970-74, 10 were from the group with average per capita incomes of $200 a year or less.[39] If this situation is to be reversed, it can be only with the co-operation of the women who are increasingly being left to manage the basic food sector almost single-handed.

References

1. Raymond Apthorpe, 'Some problems of evaluation', in Carl Gosta Widstrand (ed), *Co-operatives and rural development in East Africa*, New York, Africana Publishing Corporation, 1970.
2. Brian Scoullar, 'Pyrethrum and the highlander', *Extension Bulletin* (Department of Agriculture, Stock and Fisheries, Port Moresby) No 3, 1973.
3. Uma J Lele, *The design of rural development: lessons from Africa* (World Bank Research Publication), Baltimore, The Johns Hopkins University Press, 1975, p 26 and n15.
4. R N Coster, *Peasant farming in the Petauke and Katete areas of the Eastern Province of Northern Rhodesia*, Lusaka, Northern Rhodesia

Department of Agriculture, Agricultural Bulletin No 15, 1958, pp 11-12.

5. H A M MacLean, *Resettlement problems in the Eastern Province of Northern Rhodesia*, Lusaka, Ministry of African Agriculture, Economics and Markets Advisory Branch, 1962, pp 17-18.

6. See e g the discussion in T Scarlett Epstein, *South India: yesterday, today and tomorrow: Mysore villages revisited*, London, Macmillan, 1973, p 6.

7. Uma Lele, *op cit*, p 27.

8. *Ibid*, p 30.

9. R D Bell, Head of the Overseas Department, National Institute of Agricultural Engineering; personal communication.

10. John de Wilde, *Experiences with agricultural development in tropical Africa*, Vol 1: The Synthesis (published for the International Bank for Reconstruction and Development), Baltimore, The Johns Hopkins University Press, 1967, pp 98-100.

11. Hugh Fearn, *An African economy*, Oxford, Oxford University Press, 1961.

12. Ingrid Palmer, *The new rice in the Philippines*, Geneva, United Nations Research Institute for Social Development (UNRISD), Studies on the 'Green Revolution' No 10, 1975, p 153.

13. John de Wilde, *op cit*, pp 63-64.

14. Jane Hanger and Jon Moris, 'Women and the household economy', in Robert Chambers and Jon Moris (eds), *Mwea: an irrigated rice settlement in Kenya* (Ifo-institut für Wirtschaftsforschung, Afrika Studien 83), München, Weltforum Verlag, 1973, pp 210-11.

15. *Ibid*, pp 214-17.

16. *Ibid*, pp 229-31.

17. *Ibid*, p 234.

18. *Ibid*, p 242.

19. *Ibid*, pp 242-44.

20. *Ibid*, p 244.

21. James L Brain, 'Less than second-class ' in Nancy J Hafkin and Edna G Bay (eds), *Women in Africa: studies in social and economic change*, Stanford, Stanford University Press, 1976, pp 275-79.

22. *Ibid*, pp 275. 279.

23. Société Africaine d'Etudes de Développement (SAED), *Etude sur les besoins des femmes dans les villages de l'AVV et proposition d'un programme d'intervention*, Ouagadougou, Autorité des Amenagements des Vallées des Volta, Ministère du Développement Rural, p 56. Translation by the author.

24. *Ibid*, pp 6-33.

25. *Ibid*, p 32.

26. *Ibid*, p 6.

27. Small-Scale Irrigated Horticulture Training and Development, Chapula, *Report to the 10th meeting of the Co-ordinating Committee, 2 February 1971*.

28. Observation and interviews by the author at Chapula, September 1977. The *Interim final report*, 2nd draft, of May 1973 comments: 'An increasing amount of family labour is being used . . .' (p 14).

29. J A B Stolze, *Field report No 3, September 1972-February 1973*; also *First annual report, Growers Settlement*, p 3.

30. UNDP and FAO, *Small-scale irrigation development and training*,

Chapula: small-scale irrigated horticulture development and training, Zambia: project findings and recommendations (Terminal Report), Rome, FAO, p 31.

31. Interviews with project staff and female growers at Chapula, September 1977.

32. *Sociological survey: Chapula irrigation scheme, Chief Nkana's area,* 3 November 1972, p 2.

33. *The horticulturalist (training) semi-annual progress report covering the period 1st March to 30th September 1973,* p 3.

34. Calculated from *Nkana Growers Association: farmers' income as from 1st January − 31st May, 1975.* Grower registration numbers were matched with names on a project wallchart, *Data on Chapula farmers, trainees and settled farmers, Nkana Growers Association, 1971-74.*

35. Donald R Mickelwait, Mary Ann Riegelman and Charles F Sweet, *Women in rural development: a survey of the roles of women in Ghana, Lesotho, Kenya, Nigeria, Bolivia, Paraguay and Peru* (published in co-operation with Development Alternatives, Inc), Boulder, Colorado, Westview Press, 1976, pp 91-94.

36. John de Wilde, *op cit,* pp 191-92

37. *Ibid,* p 169.

38. Jonathan Kydd, *Family farm models and rural development planning,* unpublished paper, p 21.

39. Organization for Economic Co-operation and Development (OECD), *Development co-operation: efforts and policies of the members of the Development Assistance Committee* (1975 review), Paris, OECD, 1975, p 53.

Conclusions

The implications

Under present systems of development planning, the problems outlined in this book are rapidly deteriorating. Urgent action is needed at all levels to change the system for data collection and analysis in order to provide adequate information on women's subsistence work of all kinds, both farm and non-farm. New measures of the value of this work are also needed in order to include women in the labour force statistics, and in the national accounting systems.

What is most needed is not a series of special projects for women which perpetuate their segregation: it is vitally important that development planners who are concerned about Third World women should seek to eliminate discrimination against them in all development planning. This would mean, in the agricultural sector, that land registration would treat all working adults in a family as equal co-owners of family resources. There would have to be careful consideration of the impact on women's work of mechanization and other planned innovations. In particular, ways should be found to save women's labour and raise the returns on a given amount of time and energy expended. Depending on the needs of the area, these might include such basic facilities as wells, fuel stores or tree plantations, local markets, small grain mills, and improved tools and equipment for the cultivation of crops and the processing and cooking of food. Care should also be taken to ensure that incentives for increased production are channelled to the women as well as the men, in proportion to the contribution made by each. These measures all relate to the specific problems of rural women, the great majority; however, poor urban women suffer many of the same problems of overwork and lack of resources. It is important that discrimination against women at all levels of education and employment, formal and informal, be eliminated in urban and rural areas alike.

Postscript

This study ends where the real issues begin. There is much left to be said about Third World women and the different strategies they have adopted in the face of discriminatory development; their use of money and other resources; their potential for vastly increased productivity at all levels of the subsistence sector, as well as the production of a food surplus for sale; and their support for each other in the face of growing poverty and overwork, with forms of co-operation that could be a base for more balanced development. Many Third World researchers are currently doing vital work in various areas on the questions of what is women's real situation and what are their most urgent needs. More studies of this kind are needed. The research and its publication require support from development institutions, and the conclusions of such research need to be heard by the planners.

Index